PICKLE

AND OTHER CONDIMENT RECIPES FROM BACKYARD FARMER

Compiled by Wayne C. Whitney, Extension Horticulturalist

University of Nebraska Extension Publication CC-245 (1972)

With a new Preface by Sue Ann Gardner

AF797395

Zea Books
Lincoln, Nebraska

Compilation copyright © 1972, 2021 University of Nebraska.
Preface copyright © 2021 Sue Ann Gardner.

ISBN 978-1-60962-201-5 paperback

ISBN 978-1-60962-202-2 ebook

doi: 10.32873/unl.dc.zea.1301

Zea Books are published by the University of Nebraska–Lincoln Libraries.

Electronic (pdf) edition available online at https://digitalcommons.unl.edu/zeabook/

Print edition available from Lulu.com at https://www.lulu.com/spotlight/unllib

UNL does not discriminate based upon any protected status.
Please go to https://www.unl.edu/equity/notice-nondiscrimination

University of Nebraska Lincoln

PREFACE

The lowly pickle. It is included in an old-fashioned deli offering, a sandwich with a dill pickle spear on the side. For some, it's a nuisance, and for others, it's their favorite part of the meal. As I know many do, I fall into the latter group. So, it is my pleasure to provide a few words to preface this humble tome.

All across the United States throughout its history, wherever non-Native people moved, pickle recipes and pickle-making equipment often came with them. Pickles endure as a craft-made item in home kitchens today. In fact, there has been a resurgence of home preserving in the past several years. Pickles are an excellent entrée into this trend. The equipment can be very basic. Then you just need something to pickle, some pure water, salt, vinegar (usually but not always), and maybe one or a few other ingredients, and—after some time in a cool, dark place—voilà: *pickles*.

Homemade pickles can be better than their commercially made counterparts. In *Freezing & Canning: Prized Recipes from the Farms of America* (edited by Nell B. Nichols), pickles are aptly referred to as "The Tempting Sours." Nichols gives advice about how to have success when making pickles. First, all ingredients, equipment, surfaces, and hands should be super clean. It's best to use super-fresh, blemish-free produce; make sure the pickles are completely covered in brine; and have a tight seal on the jars. Finally, she says that pickles like to be cool or cold as they are curing rather than warm. Judith Fertig notes that the addition of a grape leaf in the jar of pickles is a secret ingredient that will help ensure crunchiness.

A fresh, crisp, sour-brined dilly cucumber pickle is a bracing and refreshing accompaniment to lunch in the fall season in northern climes, or anywhere, anytime. Slices of dill or sweet pickles are a classic topping on a burger (whether beef, bean, or any other type). But there is no reason to limit yourself to cucumber pickles. Many types of vegetables and fruits take well to pickling. Carol Stoner of *Organic Gardening and Farming* magazine once said of pickling that "Nothing else so nice ever happened to a cauliflower."

No matter the variety, pickles lend themselves to many occasions. With a stash of pickled artichokes, for instance, Susan Herrmann Loomis says that she will "slip them into sandwiches, put them on pizza, toss them with pasta, and sometimes sneak a few to eat plain, out of the jar." Pickles have a way of going with things.

Some people skip the filler and go full-on pickle. Amy Thielen describes the construction of a pickle sandwich: Fresh-made white sandwich bread hand cut into any thickness you like, a slathering of softened creamy butter, and a couple layers of thinly sliced, crisp dill cucumber pickles. Half-wrapped in waxed paper to hold it together, it's fit for a royal feast.

Depending on a person's disposition, the making of your own pickles can be thought of as a subversive act of progressive agriculture or even as a form of nonviolent protest. For one, if you have the means, you may be able to sidestep commodity produce altogether and grow your own ingredients. As Margot Ford McMillen wrote, "putting up" can be more than just a solo activity to satisfy tangible personal needs. It can promote a well-bonded community fostering "the shared optimism and confidence that gets you through winter" and can serve as a meaningful foil to corporate ag.

This small volume was compiled from recipes submitted about 50 years ago in response to a solicitation by a still-popular television show in Nebraska, *Backyard Farmer*. It's a testament to the popularity of pickles and evidence of the ingenuity that home cooks have always brought to their craft. No two recipes are alike, and each reflects a bit of the personality of the recipe submitter.

As you partake of these recipes, please be sure to use proper food-handling methods and safe food preservation techniques. Current guidelines about each of these may be accessed at USFDA Safe Food Handling (https://www.fda.gov/food/buy-store-serve-safe-food/safe-food-handling) and USDA Complete Guide to Home Canning (https://nchfp.uga.edu/publications/publications_usda.html).

I hope you enjoy many of these recipes!

Sue Ann Gardner
University Libraries, University of Nebraska–Lincoln

Selected References

Fertig, Judith. *Heartland: The Cookbook*. Andrews McMeel, 2011.

Kluger, Marilyn. *Preserving Summer's Bounty*. Evans, 1979.

Loomis, Susan Herrmann. *Farmhouse Cookbook*. Workman, 1991.

McMillen, Margot Ford. Tomato canning as protest: How a community resisted corporate farming. *Yes!* magazine, summer 2000. Available at https://www.yesmagazine.org/issue/food/2018/08/07/tomato-canning-as-protest-how-a-community-resisted-corporate-farming

Megyesi, Jennifer. *The Joy of Keeping a Root Cellar: Canning, Freezing, Drying, Smoking, and Preserving the Harvest*. Skyhorse, 2010.

Nichols, Nell B. (editor). *Freezing & Canning Cookbook: Prized Recipes from the Farms of America*. Doubleday, 1963.

Solomon, Karen. *Jam It, Pickle It, Cure It and Other Cooking Projects*. Ten Speed, 2009.

Stoner, Carol (editor). *Stocking Up: How to Preserve the Foods You Grow, Naturally*. Rodale Press, 1973.

Thielen, Amy. *The New Midwestern Table*. Clarkson Potter, 2013.

United States Department of Agriculture. Complete Guide to Home Canning [website]. Available at https://nchfp.uga.edu/publications/publications_usda.html

United States Food and Drug Administration. Safe Food Handling [website]. Available at https://www.fda.gov/food/buy-store-serve-safe-food/safe-food-handling

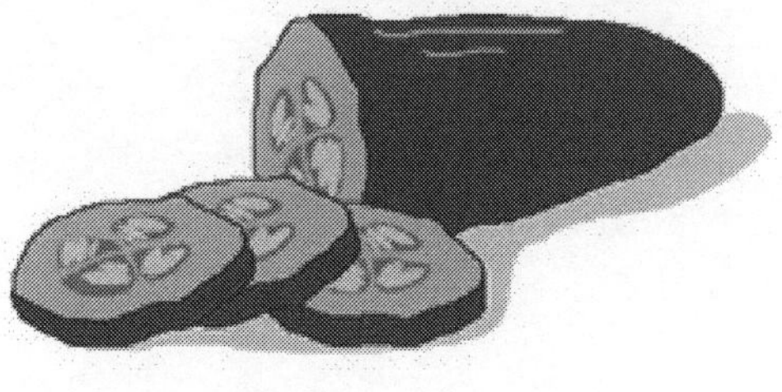

Image credit: USDA-ARS, 2017 (Public domain)

CC 245

PICKLE AND OTHER CONDIMENT RECIPES

FROM BACKYARD FARMER

Compiled by
Wayne C. Whitney
Extension Horticulturist

Introduction

Here are the favorite pickle and other condiment recipes submitted by viewers of Backyard Farmer, a television program of the Extension Service, University of Nebraska College of Agriculture. On this program, questions pertaining to the home, yard and garden are answered by specialists in the areas of Horticulture and Forestry, Entomology, Plant Pathology and Agronomy.

This publication resulted from an on-the-air request for pickle recipes. Some 536 recipes were received from interested viewers from Nebraska and surrounding states.

Recipes in this publication have not been tested or renamed. So that all recipes would be presented in the same order, ingredients have been listed at the top. No major wording alterations have been made except for listing ingredients.

These recipes have been sorted into the following groups:

Extension Service, University of Nebraska College of Agriculture Cooperating with the
U.S. Department of Agriculture and the College of Home Economics
E. F. Frolik, Dean J. L. Adams, Director

BREAD AND BUTTER PICKLES

Aristocrat Pickles

Cucumbers	1 tbsp. ground ginger
Salt	1 pt. vinegar
Alum (size of egg)	1 pt. water
1 tsp. stick cinnamon	3 lb. sugar
1 tsp. celery seed	1 tsp. whole allspice

Slice cucumbers very thin, do not peel, put in brine 8 days. Use 1 pt. salt to 1 gal. water. Drain off and cover with fresh water with alum size of egg. Simmer slowly ½ hour. Drain. Add fresh water to cover, in this add 1 tbsp. ground ginger, simmer ½ hr. Drain. Pickle solution: 1 pt. vinegar, 1 pt. water, 3 lb. sugar. Put in bag 1 tsp. stick cinamon, 1 tsp. celery seed, 1 tsp. whole allspice, add cucumbers and simmer slowly another ½ hr. Remove spice bag and seal. Secret of these pickles is slow simmering. One and one-half gallon sliced cukes makes 4 qts.

Mrs. Charles Casselman
Long Pine, Nebraska

Aristocrat Pickles

2 qt. cucumbers (sliced very thin)	2 cups sugar
6 medium onions (sliced very thin)	1 tsp. mustard seed
Salt (¼ cup to 1 qt. water)	1 tsp. celery seed
2 cups vinegar	½ tsp. pepper

Soak cukes and onions in salted water (¼ cup salt to 1 qt. water) for 2 hr. Drain well. Boil together 2 cups vinegar, 2 cups sugar, 1 tsp. mustard seed, 1 tsp. celery seed, ½ tsp. pepper. Place pickles in hot syrup and boil gently 25 min. Can.

Mrs. B. J. Maupin
Omaha, Nebraska

Best of All Pickles

12 cucumbers (good sized)	2 tsp. celery seed
6 medium onions	1 tsp. ginger
½ cup salt	¼ tsp. black pepper
2 cups sugar	1 tsp. turmeric
2 cups vinegar	

Peel and slice cucumbers about 1/8 in. thick. Prepare onions like cucumbers. Cover cucumbers and onions with salt and add water to cover. Let stand 2 hr. and drain. Let the pickling liquid (2 cups sugar, 2 cups vinegar, 2 tsp. celery seed, 1 tsp. ginger, ¼ tsp. black pepper and 1 tsp. turmeric) boil 5 min. then add onion and cucumbers. Scald through and pack into jars and seal.

Hattie Parady
Dalton, Nebraska

Bread and Butter Pickles

12 cucumbers (medium size)
6 onions (large)
2 cups sugar
1 tbsp. celery seed
1 tbsp. mustard seed
1 qt. vinegar
1 tbsp. turmeric

Slice cucumbers and onions. Let stand in salt water overnight, then pour off salt water and let stand in clear water until rest of ingredients are heated. Cook 10 min. and seal.

Mrs. Jay Collins
Alliance, Nebraska

Bread and Butter Pickles

4 qt. cucumbers (sliced)
6 onions (medium white, sliced)
2 green peppers (chopped)
3 cloves garlic
1/3 cup garlic
5 cups sugar
1½ tsp. turmeric
1½ tsp. celery seed
2 tbsp. mustard seed
3 cups cider vinegar

Do not peel cucumbers. Slice thin. Add onions, peppers and whole garlic cloves. Add salt. Cover with cracked ice, mix thoroughly. Let stand 3 hr. Drain thoroughly. Combine remaining ingredients, pour over cucumber mixture. Heat just to boil. Seal in hot sterilized jars. Makes 8 pt.

Mrs. Faye E. Haie
Beatrice, Nebraska

Bread and Butter Pickles

12 cucumbers (1½ in. circumference)
6 onions (1½ in. circumference)
2 cups sugar (white)
1 tsp. ginger
1 tsp. turmeric powder
½ tsp. black pepper
1 tsp. cornstarch
2 tsp. celery seed
2 tsp. mustard seed
2 cups vinegar

Wash, peel and slice cucumbers and onions. Let stand in salt water 2 hr. (½ cup salt). Bring sugar, cornstarch, spices and vinegar to a boil and boil 1 min. Add cucumbers and onions. Heat thoroughly and pack in jars while hot. Makes 3 pt.

Mrs. Robert M. Johnston
Emmet, Nebraska

Bread and Butter Pickles

1 gal. cucumbers (4-5 in. sliced)
8 onions (large)
4 cups vinegar
1 tsp. black pepper
1 tsp. mustard seed
2 cups sugar
1 tsp. celery seed
½ tsp. turmeric

Put cucumbers in salt water for 1 hr. Put onions, vinegar, pepper, mustard seed, sugar, celery seed, turmeric in vinegar and boil. Then put pickles in vinegar when boiling. Cook until pickles turn yellow. Can in jars.

Mrs. Harvey Wolfe
Lincoln, Nebraska

Bread and Butter Pickles

4 qt. cucumbers (medium, sliced)
6 onions (medium, sliced)
1 green pepper (sliced)
3 cloves garlic
1/3 cup salt (do not use iodized)

3 cups white vinegar (distilled)
5 cups sugar
1½ tsp. turmeric
1½ tsp. celery seed
2 tbsp. mustard seed

Choose medium-size cucumbers. Do not pare them, just wash thoroughly. Slice thin. Add sliced onion, peppers cut in narrow strips and the whole garlic cloves. Sprinkle salt over layers as you work. Now mix a tray of ice cubes through the pickles. Put another trayful on top. Let stand 3 hr. You may need more ice. It's the ice-salt mixture that makes the pickles more crisp. Drain cucumber slices thoroughly, discard the icy salt water. It's a good idea now to divide the pickles into two batches and cook in two kettles. Small batch cooking keeps the crispness and green color prized in pickles. Combine white vinegar, sugar and spices. Pour over cucumber slices. Heat just to boiling point and that is all the cooking. Seal. Let stand a month before using. Makes 8 pt. (I use a granite or stainless steel kettle to cook them in, and I use my granite roaster to let stand 3 hr.).

Mrs. Frank J. Hobbs
Lincoln, Nebraska

Bread and Butter Pickles

1 qt. cucumbers (firm, sliced)
1 cup pearl onions
1 green pepper (small, seeded, sliced)
2 tbsp. salt
Dash ground cloves

1¼ cup vinegar
1 cup sugar
½ tsp. turmeric
½ tsp. mustard seed
2 1/3 tsp. celery seed

Cover cucumbers, onions, pepper and salt with crushed ice (cover and refrigerate 3 hr. Make syrup for above from vinegar, sugar, turmeric, mustard seed, celery seed and dash ground cloves. Add vegetables, heat slowly and simmer 5 min. Put in jars and seal.

J. O. Olson
Lincoln, Nebraska

Bread and Butter Pickles

6 qt. cucumbers (sliced)
1 doz. onions (sliced)
3 qt. cider vinegar

4 cups sugar
3 tbsp. turmeric powder
½ cup mustard seed
Few cloves

Salt cucumbers and onions. Let stand overnight. No water. If too salty, wash in cold water. Mix vinegar, sugar, turmeric powder, mustard seed and cloves. Put cucumbers in large container, heat but do not boil. Soon as cucumbers become yellow, seal. Put a bay leaf on top of each jar before sealing.

Mrs. Dale A. Rogers
Lincoln, Nebraska

Bread and Butter Pickles

12 cucumbers (medium)
3 onions (medium)
Vinegar
1½ cups sugar
1 tsp. white mustard seed
1 tsp. celery seed

Slice cucumbers and onions thin (do not peel cucumbers). Soak in salt water 1 hr. Drain until dry. Use a little over 1 cup of vinegar, 1½ cups sugar, 1 tsp. white mustard seed, 1 tsp. celery seed. Boil together only 15 min. Seal in jars.

Mrs. Frank Barkus
Lyons, Nebraska

Bread and Butter Pickles

1 gal. cucumbers (sliced)
8 onions (small)
2 green peppers
½ cup salt
5 cups sugar
5 cups vinegar
1½ tsp. turmeric
1½ tsp. ground cloves
2 tbsp. mustard seed
1 tsp. celery seed

Use 1 gal. sliced cucumbers, 8 small onions, 2 green peppers, ½ cup salt. Alternate in layers, put chipped ice over this and let stand 4 hr. Drain. Heat in the liquid to the boiling point 5 cups sugar, 5 cups vinegar, 1½ tsp. turmeric, ½ tsp. ground cloves, 2 tbsp. mustard seed and 1 tsp. celery seed. Seal in jars.

Mrs. Hugh Meyers
McCook, Nebraska

Sweet Bread and Butter Pickles

1 gal. cucumbers (sliced)
8 onions (medium sliced)
1 green pepper (sliced thin)
½ cup coarse salt
5 cups sugar
5 cups vinegar
1 tbsp. mustard seed
1 tbsp. celery seed
½ tsp. turmeric powder
¼ tsp. alum

To measure, heap gallon to allow for shrinkage. About 8 medium onions sliced or use small sweet and leave them whole. 1 green pepper sliced thin. Cover with ½ cup coarse salt, stir. Let set overnight or for my convenience, for several hours. Then drain, wash salt off well and drain. Prepare 5 cups sugar, 5 cups vinegar, 1 tbsp. each mustard seed and celery seed, ½ tsp. (or more) of turmeric powder, ¼ tsp. alum. Bring to a boil and add cucumbers, onion and pepper. Simmer, do not boil, only until all well heated through, then add cut-up pimientos for color. Just heat to seal hot.

Mrs. Elmer Buchholtz
Morse Bluff, Nebraska

Bread and Butter Pickles

25 cucumbers (sliced)
½ cup salt
12 onions (sliced)
1 qt. vinegar
2 cups sugar
2 tsp. mustard seed
2 tsp. celery seed
½ tsp. ground ginger
1 tsp. turmeric powder
Some cinnamon

Soak cucumbers in ½ cup salt and water to cover for 1 hr. Drain off water, then slice 12 onions. Cook cucumbers, onions, vinegar, sugar, mustard seed, celery seed, ginger, turmeric powder and cinnamon all together and can while hot (very delicious).

Mildred Saf
Oakland, Nebraska

Bread and Butter Pickles

6 qt. cucumbers (sliced)
12 onions (medium)
3 green peppers (cup up)
½ cup salt
1 can pimiento
1 qt. vinegar
5 cups sugar
2 tbsp. mustard seed
1 tbsp. turmeric

Cover cucumbers, onions, peppers and salt with water and let stand 3 hr. Drain and add 1 small can pimientos. Heat vinegar, sugar, mustard seed, turmeric to boiling point and can (very delicious).

Mrs. Ivar Saf
Oakland, Nebraska

Bread and Butter Pickles

8 lb. cucumbers
Alum (size of walnut)
3 pt. vinegar
3½ lb. sugar
½ tsp. celery seed
½ oz. whole allspice
Cinnamon bark

Soak cucumbers in strong brine 3 days. Wash brine off and soak and wash pickles 3 days. Slice, Simmer in weak vinegar solution 28 minutes with alum size of a walnut. Pack in jars. Boil 30 min: vinegar, sugar, celery seed, whole allspice, cinnamon bark. Pour over pickles. Makes 9 pt.

Florence Estaburk
Omaha, Nebraska

Bread and Butter Pickles

2 gal. cucumbers (3-5 in.)
12 white onions (small, sliced or cubed)
4 green peppers (large, cut up)
3/4 cup coarse salt
10 cups sugar
2½ qt. white vinegar
1 tbsp. turmeric
1 tsp. cloves
1 tsp. celery seed
¼ cup mustard seed

(continued next page)

Don't peel cucumbers. Mix cucumbers, onions, peppers and salt. Let stand 3 hr. After 3 hr., drain well. Put in large kettle: sugar, vinegar, turmeric, cloves, celery seed, mustard seed. Cloves, celery and mustard seed tied into a white cloth. Let is come to boil. Put well-drained cucumbers in and bring to boil, stir, pack well in jars and seal. I usually add a pinch of alum if not sure of cucumbers. Crisp and good.

Mrs. William Chudomelka
Schuyler, Nebraska

Bread and Butter Pickles

12 cucumbers (large)
6 onions
½ cup pickling salt
3 cups sugar
½ cup water
3 cups vinegar
1 tbsp. mustard seed
1 tbsp. celery seed
1 tsp. alum

Slice cucumbers thin, soak in ½ cup pickling salt and 1 gallon cold water 3 hr. Drain off salt water but do not wash. Boil sugar, vinegar, water, mustard seed, celery seed and alum. Add cucumbers and onions. Bring to boil and can up in sterilized jars. Makes 6 pt.

Dortha McDougal Faris
Tecumseh, Nebraska

Bread and Butter Pickles

1 gal. cucumbers (1-1½ in., thin sliced)
1 qt. onions (sliced)
½ cup salt
2 trays ice cubes
5 cups sugar
5 cups vinegar
2 tbsp. mustard seed
2 tsp. celery seed
2 tsp. turmeric

Combine cucumbers, onions, salt and ice cubes. Let stand 3 hr. Drain well. In a kettle mix sugar, vinegar and spices. Bring to boil. Add drained pickles and heat thoroughly. Pack in hot, sterilized jars and seal.

Mrs. George Howard
Wahoo, Nebraska

Bread and Butter Pickles

48 cucumbers (size of quarter)
16 onions (size of walnut)
4 pt. vinegar
4 cups sugar
2 tsp. celery seed
2 tsp. mustard seed
1 tsp. ground ginger
Salt and pepper

Slice cucumbers and onions and sprinkle with salt and let stand 1 hr. Drain. Make a syrup of vinegar, sugar, celery and mustard seed, ginger, salt and pepper. Let this come to boil. Then put the drained sliced pickles in the boiling syrup. Just let come to boil. Put in hot jars and seal.

Bread and Butter Pickles

6 qt. cucumbers (medium or large)
1 doz. onions (medium)
2 green peppers (large or 4 small)
3 pt. vinegar (cider)
3 tsp. turmeric
¼ cup white mustard seed
6 cups sugar

It isn't necessary to peel the cucumbers. Slice them as thin as possible and slice onions as thin as possible. Stir in 1 cup of meat or pickling salt. Let stand 3 hr., then drain. Add the peppers, cut. Boil and pack while hot in sterilized jars and seal immediately.

Frank Fields
Shenandoah, Iowa

Bread and Butter Pickles

6 qt. cucumbers
12 onions (sliced)
4 green peppers (chopped)
1 cup salt (barrel)
3 pt. vinegar
3 tsp. turmeric
¼ cup white mustard seed
6 cups sugar

Soak cucumbers, onions, peppers in salt 3 hr. The salt will be the means of forming a juice. Drain. Pour vinegar, turmeric, white mustard seed, sugar over pickles and let come to good rolling boil. Can be used at once.

Earl May Seed & Nursery Co.
Shenandoah, Iowa

Bread and Butter Pickles

6 lb. cucumbers
2/3 cup salt
1 lb. onion (2 cups sliced)
1 small pod hot red pepper
4 cups cider vinegar
4 cups granulated or brown sugar
2 tsp. celery seed
1 tbsp. white mustard seed)
½ tsp. turmeric
1 piece ginger root (1 inch long)

Choose cucumbers about 1½ in. in diameter, 6 in. long, wash, cut off ends. Slice 1/3 in. thick, put in crockery or enamelware container. Sprinkle with salt, barely cover with cold water, stir gently. Cover. Let stand overnight. Drain 15 min. in colander. Add onions, discard stems and pepper seeds. Cut in narrow strips. Add to other ingredients. Heat slowly. Simmer 5 min. Pack hot pickles into hot sterile jars to within 1 in. of top. Fill to ½ in. of top with hot syrup. Seal at once with hot enamel-lined lids.

Mrs. J. C. Hawley
Marysvill, Kan.

Bread and Butter Pickles

25-30 cucumbers (medium)
2 large white onions
2 large sweet peppers
½ cup pickling salt
5 cups cider vinegar
5 cups sugar (2½ lb.)
2 tbsp. mustard seed
1 tsp. turmeric
½ tsp. ground cloves

(continued on next page)

Wash cucumbers and slice as thin as possible; chop onions and peppers; combine with cucumbers and salt, cover with cold water and let stand 3 hr. Drain and rinse at least once with cold water. Combine in large kettle the vinegar, sugar, mustard seed, turmeric and cloves. Bring to boil, add drained cucumbers, heat thoroughly but do not boil. Pack while hot into sterilized jars and seal at once. (Stir every once in awhile in salt water and then stir often to heat through real well before canning.)

Mrs. Geo. Maxson
North Loup, Nebraska

Company Best Pickles

10 medium cucumbers
8 cups sugar
2 tbsp. mixed pickling spices
5 tsp salt
4 cups cider vinegar

Cover whole cucumbers with boiling water. Allow to stand until next morning. Drain. Repeat this procedure on next 3 mornings. Fifth day, drain and slice in ½-in. pieces. Combine sugar, spices, salt and vinegar. Bring to boil and pour over cucumbers. Let stand 2 days. Third day, bring to boil and seal in hot, sterilized jars. Our favorite sweet pickle recipe.

Mrs. James Sterling
Lodgepole, Nebraska

Company Pickles

30 medium-size cucumbers
8 cups sugar
2 tbsp. mixed pickling spices
5 tsp. salt
4 cups vinegar

Cover whole cucumbers with boiling water, allow to stand overnight. Drain and repeat this procedure on next 3 mornings. Fifth morning, drain and slice in ½-in. pieces. I cut mine the long way. Combine vinegar, sugar, spices in bag, and salt, bring to boiling, pour over cucumbers and let stand 2 days. Third day, pack in jars, heat vinegar, pour on pickles and seal. Do not cover pickles after putting boiling water on or they will be soft.

Bev Niesen
Stapleton, Nebraska

Crispies

25 dill-size cucumbers
1 qt. coarse salt
2 tbsp. alum
1 qt. vinegar
8 cups sugar (white)
2 sticks cinnamon
1 tsp. ground mace
1 tsp. cloves

(continued on next page)

Wash cucumbers and put in brine made by combining salt with 1 gal. water. Weight down. Leave for 2 wk. and skim daily if necessary. Cut into very thin slices or if preferred, slices ½ in. thick. Cover with cold water in which 2 tbsp. alum have been dissolved and soak 24 hr. Drain and wash good. Make syrup of vinegar, sugar, cinnamon, ground mace and cloves. Put spices in a bag and bring mixture to a boil. Pour over pickles. Repeat 4 days and then put pickles in jars and can cold. Do not have to be sealed. No cooking is desired.

Mrs. Ivar G. Saf
Oakland, Nebraska

Crisp Pickle Slices

4 qt. sliced cucumbers
6 medium white onions sliced
2 green peppers--cut up
3 cloves garlic
1/3 cup salt
5 cups sugar
1 tsp. turmeric
1½tsp. celery seed
2 tbsp. mustard seed
3 cups vinegar

Mix cucumbers, onion, peppers and garlic. Add salt, cover with cracked ice. Mix good. Let stand 3 hr. Drain good. Combine remaining ingredients, pour over cucumber mixture, heat just to boiling. Seal in hot jars. Make 6-8 pt.

Mrs. Eliza Stephens
Elm Creek, Nebraska

Crisp Pickle Slices

4 qt. sliced cucumbers
6 medium white onions sliced
2 green peppers chopped
3 cloves garlic
5 cups sugar
1½ tsp. turmeric
1½ tsp. celery seed
3 cups vinegar

Do not peel cucumbers. Slice thin, add onions, peppers and whole cloves. Add salt and cover with cracked ice 3 hr. Drain well. Combine remaining ingredients, pour over cucumber mixture. Heat just to boil. Seal in hot sterilized jars. Makes about 8 pt.

Mrs. H. L. Thomas
Omaha, Nebraska

Crisp Pickle Slices

4 qt. sliced medium cucumbers
6 medium white onions sliced
2 green peppers chopped (optional)
3 cloves garlic (optional)
1/3 cup coarse-med. salt
5 cups sugar
1½ tsp. turmeric
1½ tsp. celery seed
2 tbsp. mustard seed
3 cups cider vinegar

Do not peel cucumbers, slice thin. Add onions (peppers, whole garlic, cloves). Add salt, cover with cracked ice, mix thoroughly. Let stand 3 hr., drain well. Combine remaining ingredients, pour over cucumber mixture. Heat just to boil. Seal in hot jars. Makes 8 pt.

Mr. and Mrs. Louie Koch
Stanton, Nebraska

Crisp Pickle Slices

4 qt. sliced cucumbers--medium size
6 medium white onions--sliced
1 green pepper, cut in narrow strips
1 sweet red pepper, cut in narrow strips
1/3 cup salt
3 cups white vinegar
5 cups white sugar
1½ tsp. turmeric
1½ tsp. celery
2 tsp. mustard seed

Do not pare cucumbers. Wash and slice thin. Add sliced onions and peppers. Sprinkle salt over layers as you work. Mix in a tray of ice cubes. Let stand 3 hrs. Drain. Combine vinegar, sugar and spices. Pour over cucumbers. Heat just to boiling. That is all the cooking required. Seal in pint jars. Don't use for about 1 mo.

Mrs. Ervin Shirreffs
Tilden, Nebraska

Crisp Sweet Pickle Slices

4 qt. medium cucumbers
Coarse pickling salt
1 tbsp. alum
2 tbsp. ginger
6 cups sugar
4 cups vinegar
2 cups water
1 tbsp. salt
1 tbsp. celery seed
1 tbsp. mustard seed

Slice cucumbers very thin (do not peel). Using coarse pickling salt, make a strong enough to float an egg. Cucumber slices must be completely covered. Soak 1 wk. in brine. Stir once or twice during week. Drain well. Rinse in cold water. Add 1 tbsp. of alum to enough water to cover the cucumbers and boil 5 min. Drain and rinse in clear water. Add 2 tbsp. of ginger to enough water to cover the cucumber slices. Boil again 5 min. Drain and rinse in clear water. Put cucumber slices in syrup of sugar, vinegar, water, salt, celery seed and mustard seed. Boil until clear and seal.

Mrs. Walter Swett
York, Nebraska

Easy Sliced Pickles (Delicious)

9 medium size cucumbers (4½ in.)
2 1/3 cups cider vinegar
2¼ cups sugar
½ tsp. alum
2 tbsp. salt
1 tbsp. celery seed
3¼ tsp. turmeric
3/4 tsp. mustard seed

Wash 9 medium cucumbers. Slice pickles and pour 6 cups boiling water over them. Let stand overnight. Next day mix the vinegar, sugar, alum, salt, celery seed, turmeric, mustard seed. Boil all this 5 min. Drain cucumbers and pack in glass jars (solidly). Then pour boiling hot liquid over well-filled jars. Put caps on tightly. Process in water bath for 5 min.

LeRoy and Selma Mathers
Imperial, Nebraska

Easy Sliced Pickles (Good on Hamburgers)

Cucumbers
1 tbsp. salt
1 tbsp. mustard seed
4 tbsp. sugar
Vinegar

Slice enough cucumbers of medium size to fill a pint jar (slice ¼ in. thick). Add 1 tbsp. salt, 1 tbsp. mustard seed and 4 tbsp. sugar and fill with cold vinegar and seal. No boiling necessary. They are delicious and ready to use in about 1 wk. Shake jar after sealing to dissolve the sugar and salt.

Mrs. Edwin Bader
Burr, Nebraska

Easy Sliced Sweet Cucumber Pickles

4 qt. sliced cucumbers
5 cups water
3 cups vinegar
5 cups sugar
1 tbsp. salt
3 tsp. pickling spices
1 onion

Mix all together, put sliced cucumbers in and cook 6 min. Then can and seal.

Mrs. Jay Collins
Alliance, Nebraska

Fantastic Bread and Butter Pickles

1 gal. firm cucumbers (5 in. long)
2 onions sliced
2 green peppers sliced
2 red peppers sliced
½ cup pickling salt
5 cups sugar
½ tsp. ground cloves
2 tsp. mustard seed
1 tsp. celery seed
5 cups white vinegar

Slice in thin rounds either peeled or unpeeled 1 gal. firm cucumbers around 5 in. long. Add onions, green and red peppers. Sprinkle ½ cup pickling salt over all and add 1 qt. crushed ice. Mix thoroughly. Let stand 2 hours, stir once in a while. Drain. Combine sugar, cloves, mustard seed, celery seed and vinegar. Pour over pickles in a large kettle and just bring to boil. Seal in sterilized jars.

Mrs. Harry Saum, Sr.
Grant, Nebraska

Favorite Bread and Butter Pickles

25-30 cucumbers sliced thin
2-3 onions, slivered
½ cup pickling salt
5 cups sugar
5 cups vinegar
2 tbsp. mustard seed

Mix cucumbers, onions and salt. Let set 3 hr. Drain. Heat sugar, vinegar, mustard seed. Add cukes and heat to near boiling. Seal in sterile jars.

Mrs. E. C. Volkmer
Talmage, Nebraska

Hamburger Pickles

Cucumbers (medium)
1 medium onion
Pimiento (few pieces)
2 cups sugar
1 qt. vinegar
½ cup salt (pickling)
½ tsp. powdered alum

Slice medium-size cucumbers with 1 medium onion and a few pieces of pimiento into jars. Make up syrup of vinegar, slat, sugar and alum. Bring to boil. Pour over cucumbers in jar and seal.

Mrs. Harold Kenkle
Grand Island, Nebraska

Mrs. Reed Hudson Pickles

6 qt. sliced cucumbers (unpeeled)
18 small onions sliced
2 green peppers chopped
½ cup salt
5 cups vinegar
5 cups sugar
1½ tsp. ground cloves
3 tbsp. mustard seed
3 tbsp. celery seed
½ tsp. turmeric

Mix cucumbers, onions, peppers and salt; put in bowl with ice cubes. Let stand 3 hr. and drain. Put in a large kettle, add sugar, spices; heat to boiling point. Pour into sterilized glasses and seal. Can use any size pickle.

Mrs. Reed Hudson
Lexington, Nebraska

Prize Cucumber Pickles

14 large (dill-size) cucumbers
1 qt. cider vinegar
8 cups sugar
2 tbsp. mixed pickling spices, some of red peppers removed

Wash cucumbers, cover with boiling water. Let stand 24 hr. Drain. Again cover with boiling water and let stand 24 hr. Repeat for 2 more days. Fifth day, drain/ cut cucumbers crosswise into ¼ in. slices. Combine remaining ingredients and heat to boiling. Pour hot over sliced cucumbers and let stand 24 hr. Drain off syrup, reheat to boiling, pour over cucumbers again and let stand 24 hr. Repeat 1 or 2 more days. Last day, drain. Heat syrup to boiling; add cuke slices, bring again to boiling. Pack in hot, sterilized glass jars and seal. Makes about 8 pt. Two pecks of cukes and 6 times recipe of syrup makes 38 pints of pickles. Can be canned on 8th day if desired.

Mrs. Willard M. Folsom
Lincoln, Nebraska

Sliced Cucumber Pickles

1 doz. fairly large cukes, peeled and sliced
½ doz. onions sliced
1 pt. wine
1 pt. vinegar
1 tsp turmeric
1 tsp. celery seed
1 cup sugar
1 tsp. cinnamon buds (cassia buds)
1 tsp. ginger, powdered

(Continued on next page)

Put cucumbers, peeled and sliced, sliced onions in wine or let cucumbers and onions stand in salt water 1 hr. Add vinegar, turmeric, celery seed, sugar, cinnamon buds, ginger. Scald and seal.

Mrs. J. L. Scrobbin
Oak, Nebraska

Sliced Cucumber Pickles

1 gal. cucumbers (medium)
Handful of salt
Cider vinegar
Alum (large as walnut)
1 cup horseradish root
1 tsp. mustard
1 tsp. ground clove
1 tsp. cinnamon

Put cucumbers into a jar or pail. Put enough boiling water to cover, add small handful salt. Turn them over and cover closely. Repeat this three mornings. Fourth morning, scald enough cider vinegar to cover cukes, add walnut-sized alum and a teacup finely cut horseradish root. Into a small muslin bag, put 1 tsp. mustard, 1 tsp. ground cloves and 1 tsp. cinnamon. Tie securely. Slice cucumbers ½ in. thick. Place in glass jars and pour scalding vinegar over them. Seal tightly and they will keep a good year or more.

Steve Becher
Wisner, Nebraska

Sliced Pickles

1 gal. sliced cucumbers (unpeeled)
1 cup coarse salt
1 tbsp. alum
6 cups sugar
1 tbsp. ginger
4 cups vinegar (white)
2 cups water
1 tsp. salt
1 tbsp. celery seed
1 bag mixed spices

Slice pickles as thin as possible. Soak 1 wk. in 1 gal. water to which 1 cup of coarse salt has been added. Now stir these pickles morning and night (more often if you think of it). I like to use a stone jar and have them where it's rather cool (in basement). After 1 wk., rinse and drain well. Cover again with cold water to which 1 tbsp. alum has been added. Leave 1 day. Drain and rinse well. Now place in kettle. Cover with cold water. Add 1 tbsp. ginger and boil 10 min. Rinse and drain. Now you are ready for pickling with mixture of sugar, vinegar, water, salt, celery seed and bag of mixed spices. Slowly boil 20 min. Seal hot.

Hooper, Nebraska

Sliced Pickles

25-30 cucumbers (4-5 in.)
1 qt. vinegar
8 cups sugar
2 tsp. salt
¼ cup mixed pickling spices
2-3 sticks cinnamon
Green food coloring

(continued on next page)

Cover cucumbers with boiling water 4 mornings. Fifth morning, drain. Slice pickles ½ in. thick and cover with hot syrup of vinegar, sugar, salt, mixed pickling spices (tie in little cloth) and cinnamon. Sixth morning, heat syrup to boiling point and pour over pickles again. Seventh morning, heat syrup to boil and pour over pickles again. Eighth morning, heat syrup again and drop pickles in and bring to a simmer and fill into jars and seal. Add little green coloring to syrup. Gives pickles a better color. These pickles can be kept in stone jar under vinegar.

Mabel Coleman
Merna, Nebraska

Sliced Pickles

2 doz. cucumbers (unpeeled, sliced)
1 doz. onions (sliced)
Salt
1 qt. vinegar
2 cups sugar
2 tsp. cinnamon
2 tsp. white mustard seed
2 tsp. celery seed
2 tsp. salt
2 tsp. turmeric
2 tsp. pepper

Put cucumbers and onions in separate containers. Put salt on each, about 1 cup salt for cucumbers and ½ cup salt for onions. Wash after standing 1 hr. Drain. Make brine of vinegar, sugar, cinnamon, white mustard seed, celery seed, salt, pepper and turmeric. Let come to boil. Put cucumbers and onions in this brine before you let come to boil. Seal while hot.

Mrs. Geo. Schmid
Monroe, Iowa

Sliced Pickles

Cucumbers
3 cups sugar
Spices (or onion or dill)
1/3 cup salt

Slice the pickles into a quart jar, pour the ingredients over them, place the onions, dill or spices as preferred; seal with hot lids. Keep in a cool place. Can be eaten soon.

Bessie Meduma
Weston, Nebraska

Sliced Sweet Pickles

Cucumbers
½ cup sugar
½ cup vinegar
Dill
2 tsp. salt
1 tsp. pickling spice
2 pieces garlic

For one quart jar. Put dill in bottom of jar, then sliced pickles. Add pickling spice, sugar, salt and vinegar. Shake jar until dissolved. Put jars in hot water until they change color.

Mrs. Arthur Hieb
Tripp, South Dakota

Sweet Sliced Pickles

1 gal. cucumbers
1 cup salt
2 tbsp. alum
Ginger
1 nutmeg
Few sticks cinnamon
7 cups sugar
1 qt. vinegar
1 pt. water
1 box mixed spices (to taste)

Slice and soak cucumbers 3 days in salt water. Use 1 cup salt to 1 ga. water. Rinse and soak out salt about 1 day and 1 night. Soak 24 to 48 hr. in alum water, 2 tbsp. alum to 1 gal. water. Wash well and boil ½ hr. in strong ginger water. Rinse and drain dry. Put in a stone jar. Heat sugar, vinegar, water and spices; pour over pickles for several mornings. Remove any red peppers in spice and add few sticks of cinnamon and 1 nutmeg. Pack in pint jars and seal.

Mrs. Russell Endicott
Louisville, Nebraska

Sliced Sweet Pickles

Cucumbers (dill size)
2 tbsp. salt
1 qt. cider vinegar
1 tsp. turmeric
¼ cup water
5 cups sugar
¼ tsp. alum
1 tsp. pickling spices

Use dill-size cucumbers and slice gallon crock full. Mix with 2 tbsp. salt and cover with cold water and let stand at least 2 hr. or longer. Drain, rinse once. Bring brine of vinegar, turmeric, water, sugar, alum and spices to boil and let pickles simmer about 15 min. Strain out pickling spices, put hot in jars and seal.

Mrs. J. W. Glock
Ulysses, Nebraska

Society Chips

14 dill-size cucumbers
1 qt. vinegar
8 cups sugar
2 oz. pickling spices

Wash cucumbers and cover with boiling water. Let stand 24 hr. Pour off and cover with fresh boiling water. Do this 3 mornings. Fourth morning, slice cucumbers. Heat vinegar, sugar and spices to boiling point and pour over sliced cucumbers. Let stand 24 hr. Again heat the mixture and pour over pickles. Next morning, put into jars and pour over the heated mixture and seal. Pickles are crisp. May add green color if desired.

Mrs. Joe Nelson
Ceresco, Nebraska

Society Pickles

20 dill-size cucumbers
1 qt. vinegar
8 cups sugar
1 tbsp. salt
tbsp. mixed pickling spices

(continued on next page)

Wash and cover cucumbers with boiling water. Let stand 24 hr. Pour off and cover with fresh boiling water 3 mornings. Fourth morning, slice. Heat vinegar, sugar, salt and spices. Pour on sliced cucumbers. Do this 3 mornings. Fourth morning, heat pickles and vinegar solution, add green coloring as desired. Can in jars.

Mrs. Donald Hoffman
Lincoln, Nebraska

Sweet Pickle Chips

14-15 cucumbers (4-5 in.)
8 cups sugar
1 qt. vinegar
2 tbsp. salt
2 tbsp. pickling spice
(tied in cheesecloth bag)

Wash and cover cucumbers with boiling water. Repeat 4 days. Fifth day, slice in 1/8-in. slices. Combine 8 cups sugar, 1 qt. vinegar, 2 tbsp. salt, 2 tbsp. pickling spice (tied in cheesecloth bag). Boil and pour over slices. Repeat 4 mornings. Fifth morning, simmer pickles for 2 min. Pack in jars. Add boiling syrup and seal.

Miss Melva Risse
West Point, Nebraska

Sweet Pickles--Thin Slices

Cucumbers
2 cups pickling salt
1 tbsp. alum
Green food coloring
8 cups vinegar
6 cups sugar
½ oz. celery seed
1 stick cinnamon

Slice cucumbers quite thin. Fill 2-gal. jar with sliced pickles. Dissolve 2 cups pickling salt in boiling water to cover the pickles. Let stand 8 days. Drain. Cover with boiling water. Let stand 24 hr. Drain. Cover with boiling water to which 1 tbsp. alum has been added. Let stand 24 hr. Drain. Cover with boiling water with green food coloring added if you wish to tint the pickles. Let stand 24 hr. Drain very well. Pour syrup over pickles. Heat syrup of vinegar, sugar, celery seed and cinnamon to boiling. Let stand 24 hr. Drain syrup, add 2 cups sugar, pour boiling syrup over pickles. Let stand 24 hr. Reheat liquid again, adding 2 cups sugar. Pour over pickles (12 cups of sugar will be total after adding and reheating 3 mornings). Place pickles in jars and pour over boiling liquid when you are ready to seal.

Mrs. Nettie Lovejoy
Valentine, Nebraska

Virginia Slicers

50 small cucumbers
1 tbsp. alum
2 tsp. ginger
1 tbsp. mixed spice
6 cups sugar
4 cups vinegar
2 cups water
1 tbsp. celery seed

(continued on next page)

Slice cucumbers then soak in brine that will hold an egg, 7 days. Stir well each day, drain and rinse. Boil 10 min. in water with 1 tbsp. alum. Rinse in warm water with 2 tsp. ginger. Drain and rinse again, scald until clear. Make syrup of sugar, vinegar, water, celery seed, mixed spice and salt. Cook until hot, can and seal.

Mrs. Jay Collins
Alliance, Nebraska

Virginia Slicers
(Large Cukes)

1 gal. cucumbers
½ cup salt
1 tbsp. alum
1 tbsp. ginger
2 cups water
6 cups sugar
4 cups vinegar
2 tbsp. celery seed
1 pkg. pickling spice
½ tbsp. salt

Slice cucumbers about ¼ in. Soak 5 days in brine to cover. Stir once a day. Wash 6th day. Cover with water to which 1 tbsp. alum has been added. Soak 24 hr. Wash, cover with water, add a scant tbsp. ginger. Boil 10 min. Drain and wash. Boil syrup of water, sugar, vinegar, celery seed, pickling spice, salt. When boiling, add cucumbers. Boil 20 min. or until cukes are clear and dark green. Seal in sterilized jars.

Mrs. Carl Egbers
Hooper, Nebraska

Virginia Slicers
(Don't Peel)

1 gal. cucumbers (3-4 in. long)
1 cup coarse salt
1 tbsp. alum
1 tbsp. vinegar
1 tbsp. ginger
2 cups water
4 cups vinegar
6 cups sugar
1 tbsp. celery seed
2 tbsp. mixed spices (in bag)

Slice cucumbers thin. Put sliced cukes in salt water (1 cup salt to 1 gal. water) for 1 wk., stir every day. A scum will form but stir it in, the pickles will have a tendency to clear. When week is up, drain and soak overnight in water to cover with 1 tbsp. alum added. Next morning drain and simmer 10 min. in water to cover with 1 tbsp. vinegar and 1 tbsp. ginger added. Rinse good and drain. Have hot syrup of 2 cups water, 4 cups vinegar, 6 cups sugar, 1 tbsp. celery seed, 2 tbsp. mixed spices put in bag. Simmer about 20 min. Take out spice bag and can.

Mrs. Anna Dermann
Lincoln, Nebraska

Virginia Slicers

1 gal. cucumbers
1 cup salt
1 tbsp. alum
1 tbsp. powdered ginger
6 cups sugar
2 cups water
4 cups vinegar
1 tsp. celery seed
1 tsp. salt
1 bag spices

(continued on next page)

Soak cucumbers (sliced very thin) 1 wk. in 1 gal. water and 1 cup salt. Stir every day. After 6 days soak in fresh water with 1 tbsp. alum overnight. Drain, rinse. Cook 10 min. in water with 1 tbsp. powdered ginger. Drain, rinse again. Combine sugar, water, vinegar, celery seed, salt and spices. Let come to boil. Add spices, boil 20 min.

Mrs. Arvid Johansen
Marquette, Nebraska

Virginia Slicers

1 gal. cucumbers (sliced thin)
1 cup salt
1 tsp. alum
1 tsp. ginger
6 cups sugar
4 cups white vinegar
2 cups water
1 tsp. celery seed
1 tsp. salt
2 tbsp. pickling spices (optional)

Put cucumbers in 1 gal. water and 1 cup salt for 1 wk. Stir once or twice. Rinse well. Boil 10 min. in water to cover wtih 1 tsp. alum added, rinse well again. Boil again 10 min. in water to cover with 1 tsp. ginger. Rinse and drain. Cover with syrup which has been boiled: sugar, vinegar, water, celery seed, salt, pickling spices (optional--we like them best without). Boil 10 min. Seal in sterilized pint jars.

Mrs. Willard Ivey
Raymond, Nebraska

Virginia Slicers Pickle

1 gal. cucumbers
1 cup salt
1 tsp. alum
1 tsp. ginger
6 cups sugar
4 cups vinegar
1 tsp. celery seed
Mixed spices (in a bag)

Slice cucumbers as thin as possible. Soak in 1 gal. water and 1 cup salt for 1 wk. Stir every day. Drain and rinse--boil 10 min. in water with 1 tsp alum. Drain and rinse again. Boil 10 min. in water with 1 tsp. ginger. Drain and rinse. Bring syrup of sugar, vinegar, celery seed and spices to boil. Add drained pickles and boil 10 min. Seal.

Mrs. Wm. J. Murphy
O'Neill, Nebraska

Virginia Slicers (Sweet Pickles)

1 gal. cucumbers
1 cup coarse salt
1 tbsp. powdered alum
1 tbsp. ginger
6 cups sugar
2 cups water
4 cups vinegar
1 tbsp. table salt
2 tbsp. mixed pickling spices (tied in a cheesecloth bag)

(continued on next page)

Wash but do not peel cucumbers. Slice cucumbers about ½ in. thick. Put in a stone jar and cover with 1 cup coarse salt and 1 gal. cold water. Let set 1 wk. Stir every day. A scum will form on top but stir it in. These will smell but won't spoil and will clear some by the end of the week. When week is up, drain, wash and rinse well. Put on more cold water with 1 tbsp. powdered alum in it. Let stand 2 days. Drain and rinse well. Put on more cold water to cover with 1 tbsp. ginger. Boil 10 min. Drain and rinse again. Have the following ready: sugar, water, vinegar, table salt, mixed spices. Put the cucumber slices into this syrup and boil 20 min. Seal in sterilized jars.

Mrs. Marion Spangler
Weeping Water, Nebraska

Virginia Slicers

1 gal. sliced cucumbers (unpeeled)
1 cup salt
1 tbsp. alum
1 tbsp. ginger
8 cups sugar
1 tsp. celery seed
2 2/3 cups water
5 1/3 cups vinegar
½ box pickling spice
1 tbsp. salt

Soak cucumbers in brine made of 1 gal. water and 1 cup salt. Stir every day for 5 days. Soak in 1 gal. water, mixed with alum on 6th day. Drain and rinse. Simmer drained slices in 1 gal. water and the ginger 20 min. Drain and rinse. Tie pickling spices in a bag. Combine sugar, water, vinegar and spices, bring to boil and add cucumbers. Boil 20 min. or until slices are transparent and clear. Seal. These are almost candied.

Mrs. Neil Chandler
Imperial, Nebraska

CHUNK PICKLES

Best Chunk Pickles

25 good-sized cucumbers
1 qt. salt
Alum
1 qt. vinegar
2 qt. sugar
2 sticks whole cinnamon
Whole colves

Wash cucumbers. Slice ½ in. thick and soak 2 wk. in brine of 1 qt. salt and 4 qt. water in stone jar. Drain and wash. Add alum the size of 3 walnuts and cover with water. Let stand 24 hr. Drain and wash. Boil vinegar, sugar, cinnamon and cloves. Pour over pickles. Second morning, drain and reheat. Pour over pickles. Repeat 3rd morning. Fourth morning, add ½ cup sugar to mixture and heat again. Pack pickles in jars and seal hot.

Mrs. H. Mickle
Waverly, Nebraska

Chunk Pickles

1 ga. medium cukes
1 cup salt
¼ tsp. alum
3 pt. vinegar
4 cups sugar
1 stick cinnamon
½ tsp. whole cloves
½ tsp. allspice
1 tbsp. mustard seed

For 1 gal. medium cukes, make brine of 1 cup salt and 1 gal. water. Let stand 3 days. Drain, let stand in cold water 2 days. Sixth day, drain and cover wit ice water with ¼ tsp. alum. Seventh day, drain and cut into chunks and cook in weak vinegar (2 pt. vinegar to 2 qt. water) about 15 min. Make syrup of 4 cups sugar, 1 pt. vinegar, cinnamon, cloves, allspice and mustard seed. Pour over pickles. Reheat syrup 2 mornings. Third morning, seal while hot. Yields 5 pt.

Mrs. Ethel Borgman
Lincoln, Nebraska

Chunk Pickles

25 dill-size cucumbers
1 pt. salt
Alum size of 2 walnuts
1 qt. vinegar
2 qt. sugar
2 sticks cinnamon
1 tsp. mace
1 tsp. whole cloves

Soak cucumbers in 1 pt. salt to 1 gal. water 2 wk. in stone jar. Wash, drain, cover with cold water overnight with alum. Wash, drain and cut in chunks. Make syrup of vinegar, sugar, cinnamon, mace and cloves, let syrup come to boil, pour over pickles while hot. Do this 4 days. Then can and seal.

Mrs. A. E. Heins
Columbus, Nebraska

Chunk Pickles

7 lb. large green cucumbers
Vinegar
1 oz. celery seed
1 oz. cinnamon bark
2 lb. sugar
1 oz. whole allspice
1 tsp. salt

Cut cucumbers in chunks. Heat them in weak vinegar 2 hr. (Do not boil). Put pickles in jars and pour over them the following while hot: 3 pt. vinegar, celery seed, cinnamon bark, sugar and allspice. Add salt and seal.

Mrs. George Schmid
Monroe, Nebraska

Crispy Chunks

8 cukes, 5 in.
2 cups vinegar
4 cups water
2 tsp. alum
4 cups sugar
2 cups white vinegar
1 tbsp. pickling spices

Cover cukes with brine (½ cup coarse-medium pickling salt to 1 qt. water). Let stand 3 days. Drain. Cover with cold water. Let stand 3 days, draining and covering with fresh cold water each day. Cut in ½-in. chunks. (I cut them about 1 in. thick.) Combine 2 cups vinegar, 4 cups water and alum. Heat to boiling, pour over chunks. Let stand 2 days. Drain. Combine sugar, 2 cups white vinegar and spices. Heat to boiling, pour over pickles. Let stand 1 day. Pour off liquid, heat to boiling and pour over pickles. Do this 3 times, daily. The last time, put cukes in jars. Heat liquid and pour over them in jars. Makes 4 pt.

Mrs. Sam Lane
Council Bluffs, Iowa

Sweet Chunk Dill Pickles

Cucumbers
Salt
Ice cubes
1 cup vinegar
1 cup sugar
1 tsp. mustard seed
Dill
1 tsp. celery seed

Wash and cut cucumbers in chunks. Sprinkle with salt well, put ice cubes on top. Let stand 5 hr. Wash and drain. Boil vinegar and sugar. Drop cucumbers in boiling liquid, let come to boil. Put in quart jars. Add mustard seed, on top, dill, celery seed, on top. Seal hot. (2 qt. vinegar and other ingredients in proportion make 6 qt. pickles. This is nice when pickles are picked from small garden as you can make 1 or 2 qt. at a time.)

Mrs. J. L. Scroggin
Oak, Nebraska

Sweet Chunk Pickles

Cucumbers
3 tsp. alum
1 gal. water
10 cups sugar
1 qt. vinegar
2 bags pickling spice
Salt

Soak whole cucumbers in salt water (strong enough to float an egg) 7 days. Put fresh water on them 3 mornings. After 7th day, wash pickles and jar also. Slice about 1 in. and boil in alum water about 50 min. or until centers change color. Rinse several times in cold water. Put back in large jar, make a syrup of 10 cups sugar to 1 qt. vinegar and a little water, 2 bags of pickling spice (size of egg). Boil syrup 5 min. Pour hot over pickles. Heat syrup 3 mornings and pour over pickles. Add 1 cup sugar each morning to suit taste. Put spices in bags and boil with syrup each morning. Fourth morning, pack pickles in jars and pour hot syrup on them. Seal.

Mrs. Joe Nelson
Ceresco, Nebraska

Sweet Chunk Pickles
(My Favorite)

Cucumbers
1 cup salt
Vinegar
6 cups sugar
1 stick cinnamon
1 tsp. mustard seed
½ tsp. whole cloves
½ tsp. whole allspice

Put cucumbers in brine 3 days, using 1 cup salt for each gal. water it takes to cover cucumbers nicely. Turn a plate over pickles in 3- or 5-gal. crock, whichever it happens to be. Weight plate down with qt. jar full of water. This keeps cucumbers in brine. Take cucumbers from brine and replace with clear cold water 3 days. Seventh day, cut cucumbers in chunks. (I cut them in half lengthwise. If they are too large, I cut them in half crosswise. If larger, they can be cut in lengths of about 1¼ in. across the pickle.) Make a weak vinegar solution using 2 cups of vinegar to 1 gal. water. Bring to boil. Drop in cucumber chunks and just bring to boil. Set aside right away and let cucumbers heat through. Let stand in vinegar solution 3 days. Drain. Make syrup of sugar, 2 cups vinegar, cinnamon, mustard seed, cloves and allspice. To keep spices from darkening the pickles, tie them in a piece of cheesecloth or part of a lady's silk hose (clean) will do as well. Pour syrup over cukes 3 mornings, heating vinegar each morning. You may have to triple ingredients as you must have enough to cover cucumbers nicely. Third morning, put cukes in sterile jars, heat vinegar and seal while hot. This is my favorite sweet pickle recipe. Very good when kept in icebox and are cold.

Addie Bradley
Columbus, Nebraska

Sweet Chunk Pickles

7 lb. cukes
Salt
4 pt. vinegar
Alum
3 lb. white sugar
Handful whole allspice

(continued on next page)

Soak cukes 3 days in strong salt water. Drain and soak 2 days in clear water. Drain and weigh to 7 lb. Take 1 pt. vinegar, 3 pt. water and alum size of walnut. Let simmer 2 hr. on low flame. Take from this and put in jars. Make a pickle juice, 3 pt vinegar, 3 lb. white sugar and allspice. Bring to boil and pour over and seal.

Robert M. Johnston
Emmet, Nebraska

Sweet Chunk Pickles

7 lb. cucumbers
Salt
Alum
2½ lb. sugar
1 oz. cinnamon bark
1 oz. celery seed
1 oz. allspice
3 pt. vinegar

Soak cucumbers 3 days in salt water that will float an egg. Change the water each day. Simmer in alum water 3 hr. Boil sugar, cinnamon bark, celery seed, allspice and vinegar 5 min. Pour over drained hot pickles and seal while hot.

E. H. Williams
Norfolk, Nebraska

Sweet Chunk Pickles (Tested)

1 gal. large or medium cucumbers
3/4 cup pickling salt
Vinegar
½ tsp. alum
4 tbsp. mustard seed
1 tbsp. whole cloves
1 tsp. turmeric
Sugar

Cut cukes into 1½-in. pieces. Soak in salt water 3 days. Salt water is made of 2 gal. water and 3/4 cup pickling salt (use ½ of thic recipe for 1 gal. pickles). When they have soaked 3 days, drain and cook in 1 cup vinegar, 1 cup water and ½ tsp. alum. Drain. Boil 2 qt. and 1 cup vinegar, mustard seed, whole cloves, turmeric, 2 qt. and 1 cup sugar (in vinegar). Put pickles into syrup, heat thoroughly and seal.

Mrs. Walter Jeary
Seward, Nebraska

Virginia Chunk Sweet Pickles

Cucumbers
2 cups salt
Alum
6 cups vinegar
Sugar
1/3 cup pickling spice
1 tbsp. celery seed

Use 75 cukes 4 or 5 in. long (or 2 gal. small ones). Make brine of 2 cups salt to 1 gal. water. Boil and pour over pickles. Let stand 1 wk. Skim if necessary. Drain and cut. For 3 mornings, make solution of 1 gal. water and 1 tbsp. alum.

(continued on next page)

Pour over pickles. Change each morning and use fresh solution of water and alum, boiling hot. Fourth morning, heat 6 cups vinegar and 5 cups sugar, 1/3 cup pickling spice, 1 tbsp. celery seed and bring to boiling point. Pour over pickles. Fifth morning, drain off syrup, add 2 cups sugar. Heat and pour over pickles. Sixth morning, drain off syrup and add 1 cup sugar. Heat and pour over pickles that have been packed into sterilized jars and seal at once. Use crock for making these pickles.

Mrs. J. G. Porter
Lincoln, Nebraska

Virginia Chunk Sweet Pickles

Cucumbers	6 cups vinegar
Salt	1/3 cup pickling spice
Alum	1 tbsp. celery seed

Use 75 cucumbers 4 or 5 in. long (2 gal. small ones). Or use what you have. Make brine of 2 cups salt to 1 gal. water. Boil and pour over cucumbers boiling hot. Let stand 1 wk. In hot weather, skim daily. Drain and cut in chunks. For next 3 mornings, make boiling hot solution of 1 gal. water and 1 tbsp. powdered alum and pour over pickles. Make this fresh hot bath 3 mornings. Fourth morning, drain from alum water and heat 6 cups vinegar, 5 cups sugar, 1/3 cup pickling spice and 1 tbsp. celery seed to boiling point and pour over pickles. Fifth morning, drain liquid off and add 2 more cups sugar. Heat again to boiling point and pour over pickles. Sixth morning, drain liquid, add 1 cup sugar, heat, pack pickles into sterilized jars and pour liquid over pickles until jars are filled. Seal while hot.

Florence Benson
Omaha, Nebraska

Virginia Chunk Sweet Pickles

Cucumbers	Sugar
Salt	1/3 cup pickling spice
Alum	1 tbsp. celery seed
6 cups vinegar	

Use 75 cucumbers 4 or 5 in. long (or 2 gal. small ones or use what you have). Most any nice solid cucumber makes a nice pickle. Brush and clean in water. Make brine of 2 cups salt to 1 gal. water, boil and pour over cucumbers boiling hot. Weigh down to keep under brine. Let stand 1 wk. In hot weather, skim daily. Drain and cut in chunks. For next 3 mornings make boiling hot solution of 1 tbsp. powdered alum and pour over pickles. Make this fresh hot bath 3 mornings. Fourth morning, drain and discard alum water. Heat 6 cups vinegar, 5 cups sugar, pickling spice and celery seed to boiling point and pour over pickles. Sixth morning, drain liquid. Add one cup sugar, heat. Pack pickles into sterilized jars and fill to within ½ in. of top with boiling liquid. Seal at once.

Mrs. William R. Wessel
Roca, Nebraska

20-Minute Chunk Pickles

4 qt. cukes, sliced or chopped
5 cups vinegar
5 cups sugar
3 cups water
3 tsp. salt
1 tsp. black pepper
1 tsp. cloves
1 tsp. allspice
1 tsp. cinnamon
2 medium onions, finely cut

Mix vinegar, sugar, water, salt, black pepper, cloves, allspice and cinnamon; boil. Put spices in bag. After mixture begins to boil, add cucumbers and onions. Continue to boil 5 min. Can in sterilized jars.

Earl May Seed & Nursery
Shenandoah, Iowa

12-Day Chunk Pickles

Cucumbers
Salt
Vinegar
3 lb. sugar
1 stick cinnamon
1 tbsp. mustard seed
½ tsp. whole cloves
½ tsp. whole allspice

Select medium cucumbers and allow to stand in brine (1 cup salt to 1 gal. water) 3 days. Drain and cover with cold water 3 days. Seventh day, cut in chunks and cook in weak vinegar until tender. Let stand 3 days. Make syrup of sugar, 1 pt. vinegar, cinnamon, mustard seed, cloves and allspice. Pour hot syrup over pickles 3 mornings, then seal.

Dorothy & Emmet Erickson
Oakland, Nebraska

12-Day Chunk Pickles

Cucumbers
Salt
Vinegar
Sugar
Cinnamon
Mustard seed
Whole cloves
Whole allspice

Use 1 cup salt and 1 gal. water for brine. Soak cucumbers 3 days. Let stand 3 days in clear water. Chunk the pickles, cool until tender in weak vinegar solution (1 cup vinegar to 2 cups water). Let stand 3 days. Drain. Heat syrup made of 3 lb. (6 cups) sugar, 1 pt. vinegar, 1 stick cinnamon, 1 tbsp. mustard seed, ½ tsp. whole cloves and ½ tsp. whole allspice and pour over pickles 3 mornings. (Tie spices into a bag.)

Mrs. Nettie Lovejoy
Valentine, Nebraska

12-Day Chunk Sweet Pickles

Cucumbers
Salt
Vinegar
4 cups sugar
1 stick cinnamon
1 tbsp. white mustard seed
Allspice
Cloves

(continued on next page)

Make brine of 1 cup salt to 1 gal. water. Put in as many pickles as brine will cover. Allow to stand in brine 3 days then drain. Let stand in clear water 3 days. Seventh day, cut in chunks and cook in 1 cup vinegar to 2 cups water until tender. Let stand in this liquid 3 days, the drain. Make syrup of sugar, 1 pt. vinegar, small amount of water, cinnamon, mustard seed, allspice and cloves. Bring to boil and pour over pickles. Repeat 3 times, then pour hot syrup over pickles in jars. Seal.

Hattie Parady
Dalton, Nebraska

CRYSTAL PICKLES

Crystal Chip Pickles

14 medium-size cucumbers
8 cups sugar
4 cups vinegar
2 tbsp. pickling spices
1 heaping tbsp. salt

Pour hot water over cucumbers 3 days. Slice ¼ in. thick now. Make syrup of sugar, vinegar, pickling spices and salt. Pour syrup over cucumbers 3 mornings. Reheat 3 mornings. Seal.

Mrs. Otto Schmidt
Arlington, Nebraska

Crystal Pickles

25 medium-size cucumbers
(or dill size)
Salt
Crystal alum size of walnut
1 qt. vinegar
1½ qt. sugar
2 sticks cinnamon
1 tbsp. whole cloves

Make brine strong enough to float an egg. Enough to cover cucumbers. Let stand 2 wk. Drain. Wash and split or cut cucumbers in chunks and put in clear water to which has been added the alum. Let soak 2 days, changing water each day. Drain. Bring to boil vinegar, sugar, cinnamon and cloves (cinnamon and cloves should be tied in sack). Pour on sliced or cut cucumbers. Repeat by bringing syrup to boil 3 mornings. If desired, add a little more sugar every morning. Fourth morning, pack cucumbers in jars and pour over hot syrup and seal.

Mrs. Henry Wendt
Elmwood, Nebraska

Crystal Pickles

4 lb. cucumbers (dill size)
Salt
2 tbsp. alum
1 qt. vinegar
2 qt. sugar
2 sticks cinnamon
1 tsp. whole cloves

Put cucumbers in brine--1 qt. coarse salt to 1 gal. water--2 wk. Put plate over to weight down. Keep in cool place. Drain, wash, cut in ¼-in. slices, cover with cold water and alum. Soak 24 hr. Drain, wash. Put in large container in syrup made of vinegar, sugar (plus cinnamon and cloves in a bag). Bring to boil and pour over. Repeat 4 days. Seal. Syrup covers 1 gal.

Mrs. Nellie G. Dwinell
Lincoln, Nebraska

Crystal Pickles

25 dill-size cucumbers
1 qt. coarse salt
2 tsp. powdered alum
1 qt. vinegar
2 qt. sugar
2 sticks cinnamon
1 tsp. ground mace
1 tsp. whole cloves

Wash cucumbers, put in brine made with salt and 1 gal. water. Use stone jar or enamel kettle. Cover with a plate and weight down (I use a 2-qt. jar half-filled with water) so all cukes are below the brine. Cover with clean cloth. Skim daily if necessary. Leave 2 wk. Drain, wash. Cut in slices about ½ in. thick. Cover with cold water and alum. Soak 24 hr. Drain, wash. Make syrup of vinegar, sugar, cinnamon, mace and cloves. Put spices in bag. Bring to boil, pour over pickles. Repeat 4 days. This means drain pickling syrup, repeat to boiling and again pour over pickles. This syrup covers about 1 gal. of chunks. Good-size cucumbers may be used but be sure large seeds are not formed. Cucumbers must not be older than 24 hr. or they will not make crisp pickles. You may add fresh cucumbers to the brine from time to time but allow 2 wk. for the last added. These may be allowed to remain in brine a long time but need more soaking to remove salt. Put pickles in jars when process is complete. Can cold. They are good at once but improve on standing about 6 wk. Then you will have the most delicious pickles you ever ate. They will be crisp, wonderful tasting and have a lovely clear dark-green color.

Mrs. Glen Miller
Louisville, Nebraska

Crystal Pickles

25 dill-size cucumbers
Salt
Alum
1 qt. vinegar
2 qt. sugar
1 tsp. mace
1 tsp. cloves

Put cukes in strong brine 2 wk. Make brine of 1 qt. salt to 6 qt. water. Stir once in a while. After 2 wk., slice and cover with cold water containing alum size of walnut. Let stand overnight. In morning, wash, drain. Mix vinegar, sugar, mace and cloves, let come to boil and pour over pickles while hot. Let stand overnight. Repeat 4 days. These will keep in open jars.

Mrs. Geo. Schmid
Monroe, Nebraska

Crystal Pickles
("Adella Shoemaker")

25 dill-size cucumbers
Coarse salt
2 tbsp. powdered alum
1 qt. vinegar
2 qt. sugar
2 stick cinnamon
1 tsp. ground mace
1 tsp. whole cloves

(continued on next page)

Wash cucumbers, put in brine made with 1 qt. coarse salt and 1 gal. water. Use a stone jar. Cover with a plate and weigh down. (I use a 2-qt. jar half-filled with water so all cukes are below brine.) Cover with clean cloth and skim daily if necessary. Leave 2 wk. Drain, wash. Cut in slices about ½ in. thick. Cover with cold water and alum. Soak 24 hr. Drain, wash. Make syrup of vinegar, sugar, cinnamon, mace and cloves. Put spices in bag. Bring to boil, pour over pickles. Repreat 4 days. This means to drain pickling syrup, repeat to boiling and again pour over pickles. This syrup covers about 1 gal. of chunks. Good-size cucumbers may be used but be sure large seeds are not formed. Cucumbers must not be older than 24 hr. or they will not make crisp pickles. You may add fresh cucumbers to brine from time to time but allow 2 wk. for last added. These may be allowed to remain in brine for longer time but need more soaking to remove salt. Put pickles in jars when process is complete.

Mrs. James Rader
Nebraska City, Nebraska

Crystal Pickles

25-30 cucumbers
Salt
2 tbsp. alum
1 qt. vinegar
2 qt. sugar
2 sticks cinnamon
1 tsp. whole cloves
1 tsp. mace (in bag)

Cover cucumbers with brine made of 1 qt. salt to 1 gal. water. Let stand 2 wk. Weigh down. Drain, wash, cut in thick slices ¼ to ½ in. Cover with cold water to which alum has been added. Boil vinegar, sugar, cinnamon, cloves and mace. Heat, pour over pickles 5 mornings. Let cook and put in jars.

Mrs. Harvey Haecker
Pickrell, Nebraska

Crystal Pickles

25 dill-size cukes
Salt
2 tbsp. alum
1 qt. vinegar
2 qt. sugar
2 sticks cinnamon
1 tsp. mace
1 tsp. whole cloves

Put cukes in brine made of 1 qt. salt to 4 qt. water. Weight down and let stand 2 wk. Remove and wash, slice into chunks. Cover with cold water plus alum and soak overnight. Wash, drain. Make syrup of vinegar, sugar, cinnamon, mace and cloves. Bring to boil and pour over pickles. Repeat 4 times. Will keep in open jar or can put into jars the last day and pour syrup over hot. Can start with fewer cukes and add as you pick more but count time in brine after adding last ones. No harm done if allowed in brine a little over time.

Mrs. A. W. Tatum
Riverdale, Nebraska

Crystal Pickles

25 dill-size cukes
1 qt. pickling salt
2 tbsp. powdered alum
1 qt. vinegar
2 qt. sugar
2 sticks cinnamon
1 tsp. whole cloves

Wash cucumbers and put in a stone jar in brine made of salt and 1 gal. water. Leave in brine 2 wk. Skim every day. Take cucumbers out, drain, wash. Cut in chunks or slice about ¼ to ½ in. thick. Cover with cold water to which you have added alum. Let stand 24 hr. Wash, drain. Make syrup of vinegar, sugar, cinnamon and cloves. Put spices in bag (cloth) and bring to boil. Pour over pickles. Fourth day, put pickles in jars and cover with hot syrup and seal.

Mrs. C. H. Propst
Seward, Nebraska

Crystal Pickles

25 dill-size cucumbers
1 qt. pickling salt
2 tbsp. alum
1 qt. cider vinegar
2 qt. sugar (granulated)
2 sticks cinnamon
1 tsp. whole cloves
1 tsp. ground mace

Wash cucumbers not over 24 hr. old. Place in stone jar. Pour over cucumbers a brine made of salt and 1 gal. cold water. Weight down with clean plate and qt. jar filled with water (lid on, too). Leave in brine 2 wk.--cloth over jar. Drain, wash well, cut in slices. Cover with cold water and alum 24 hr., then drain, wash, place back in clean stone jar. Pour over cucumbers a syrup, boiling hot, made of vinegar (including small amount green color), sugar, cinnamon, cloves and mace (spices in bag). Drain each following 4 mornings. Five in all. Process only in stone jar. These can be canned in sterilized fruit jars--syrup heated and put over them. Or canned in large glass containers having lids.

Mrs. W. H. Howarth
Tecumseh, Nebraska

Crystal Pickles
(Best Pickle Ever Made)

25 fresh dill-size cukes
1 qt. pickling salt
2 tsp. powdered alum
1 qt. vinegar
2 qt. sugar
2 stick cinnamon
1 tsp. mace
1 tsp. whole cloves

(continued on next page)

Stir pickling salt into 1 gal. water. Wash fresh cukes, put in stone jar. Pour brine over cukes. Weigh down, skim scum from brine each day for 2 wk. Keep in cool place. Wash and drain. Cut in desirable chunkes. Add alum with enough cold water to cover chunks in crock. Let stand in alum water 24 hr. Wash in fresh water, drain. Boil pickles syrup made of vinegar, sugar, cinnamon, mace and cloves. (Put spice in bag). Pour over cukes in crock. For next 4 days, pour syrup off cukes. bring to boil and pour over cukes again. After 4 days, put cukes in sterilized jars. Pour syrup over cukes. Process jars 8-10 min. in hot water bath or simmer 20 min. in open kettle, pack in sterilized jars hot. Seal.

Mrs. Mildred Beyer
Wahoo, Nebraska

Crystal Pickles

25 dill-size cukes (not cut)
1 qt. salt
2 tbsp. alum
1 qt. vinegar
2 qt. sugar
2 tbsp. cinnamon
(or 3-4 sticks cinnamon)
1 tsp. whole cloves
1 tsp. mace

Use 24-hr. old cukes. Put into stone jar, cover with brine made of salt and 1 gal. water. Use a plate to weigh down. Skim if necessary. Let stand 2 wk. Remove and wash well. Cut in chunks ½ to 1 in. thick. Cover with cold water with alum. Let stand 24 hr. Drain, wash. Make syrup of vinegar, sugar, cinnamon, cloves and mace (spices in bag), boil and pour over pickles. Second morning, pour off syrup and bring to boil. Pour over cukes again. Third morning, repeat. Fourth morning, bring to bring syrup and cukes to boil. Seal in jars.

Mrs. Murl Smith
Defiance, Iowa

Delicious Crystal Pickles

25 dill-size cucumbers
1 qt. salt
2 tsp. powdered alum
1 qt. vinegar
2 qt. sugar
2 sticks cinnamon
1 tsp. mace
1 tsp. whole cloves

Wash cucumbers and put in brine made of salt and 1 gal. cold water. Use stone jar of enamel kettle. Cover with plate and weigh down. Skim daily if necessary. Leave 2 wk. in brine. Drain, wash. Cut in slices ½ in. thick. Cover with cold water and alum. Soak 24 hr. Drain, wash. Make syrup of vinegar, sugar, cinnamon, mace and cloves (put spices in bag). Bring to boil and pour over pickles. Heat syrup and repeat 4 days. When process is complete, put in jars, can cold.

Mrs. Mary P. Thomas
Pawnee City, Nebraska

Glazed Crystal Pickles

Cucumbers
Salt
3 cups vinegar
Sugar
Pickling spices

Make a salt brine strong enough to float an egg (1 lb. salt to 9 pt. water). Soak whole cucumbers in brine 3 days. Drain off brine and soak in clear fresh water 3 days. Change water each day. Drain. Make a solution of 1 cup vinegar and 1 scant tsp. alum to 4 cups water. Make enough to cover pickles. Soak 3 days in this solution. Drain well, then cube cucumbers. Make syrup of 2 cups sugar, 2 cups vinegar and 1 tbsp. pickling spices. Boil syrup and pour hot over cucumbers. Each day for succeeding 2 days pour off syrup, add 2 cups more sugar, bring to boil, pour over pickles. (Adding all the sugar at one time would shrink the pickles.)
Third day, pack pickles in clean, hot sterilized jars and seal.

Mrs. Clara Loberg
Norfolk, Nebraska

Sweet Crystal Pickles

25 cukes
1 qt. pickle salt
2 tbsp. alum
1 qt. vinegar
2 qt. sugar
1 tsp. ground mace
1 tsp. whole cloves

Put cukes in brine made of salt and 1 gal. water. Put in jar and weight down and let set 2 wk. Skim when necessary. Drain, wash. Slice ½ in. thick. Cover with cold water plus alum. Soak 24 hr. Drain, wash. Add syrup, boiling hot, made of vinegar, sugar, mace and cloves (spices in bag). Boil. Repeat this draining, reheating and repouring process 3 more more times--4 times in all--once each day. Pack into clean jars and cover. (Not necessary to seal tightly, such as in other processing.)

Mrs. E. C. Volkmer
Talmage, Nebraska

CURRY PICKLES

Curry Pickles

3 lb. 4-to-5-in. cucumbers
1 2/3 cups white veingar
1 cup granulated sugar
1½ tsp. curry powder
2 tbsp. mustard seed
1½ tsp. celery seed
1 cup water
2 tbsp. salt

Wash cucumbers thoroughly and cut into chunks. Combine remaining ingredients. Heat to boiling. Add cucumbers. Heat just to boiling point, then simmer while quickly packing one hot, sterilized jar at a time. Fill to 1/8 in. from top of jar. Be sure vinegar solution covers vegetables. Seal each jar before filling the next. Makes four pints.

Mrs. Val Kuska
Lincoln, Nebraska

Curry Pickles

12 medium cucumbers
Salt
½ tsp. curry powder
1¼ cup vinegar
1 cup sugar
1 tsp. mustard seed
1 tsp. turmeric
1 tsp. celery seed

Peel and slice cucumbers. Make brine of 4 cups water and 1 tbsp. salt. Pour over sliced cucumbers. Let stand 5 hr. or less. Rinse well in clear water. Combine curry powder, vinegar, sugar, mustard seed, turmeric and celery seed and bring to boil. Add slices and bring to boil. Boil 10 min., stirring gently. Seal in jars.

E. H. Williams
Norfolk, Nebraska

Curry Pickles

24 medium cukes (cubed or sliced)
½ cup salt
8 cups water
2 cups vinegar
2½ cups sugar
¼ cup mustard seed
1 tbsp. celery seed
1 tsp. curry powder

Wash cukes. Drain, combine salt and 8 cups water. Pour over cukes. Let stand 5 hr. Drain, rinse thoroughly. Combine remaining ingredients. Heat to boiling, pour over cucumbers, heat to boiling and pack in sterilized jars and seal at once.

Mrs. E. L. Lindberg
Omaha, Nebraska

Curry Swedish Pickles

50 small cucumbers
Salt
1 qt. vinegar
1 qt. water
6 cups sugar
2 tbsp. mustard seed
1 tbsp. celery seed
1 tsp. curry powder

(continued on next page)

Soak cucumbers 3 hr. in salt brine to cover (2 cups non-iodized salt to 1 gal. water). Drain and cut pickles into 1-in. chunks. Make syrup of vinegar, water, sugar, mustard seed, celery seed, curry powder and 1 tbsp. salt. Pour cucumbers in syrup and bring to boil (but do not boil). Pack in hot sterilized jars and seal. This is a delicious rather unusual pickle given my family many, many years ago by a Swedish friend.

Mrs. Louis Gross
Fremont, Nebraska

Easy Curry Cucumbers

24 medium cucumbers
½ cup salt
8 cups water
1 tbsp. curry powder
2 cups vinegar
2½ cups sugar
¼ tsp. mustard seed
1 tbsp. celery seed

Combine salt and water and soak cucumbers in this 5 hr. Drain and rinse. Combine curry powder, vinegar, sugar, mustard seed and celery seed, heat to boiling, pour over cucumbers, heat to boiling and pack in jars. Seal. Makes 3 qt.

Mrs. C. M. Stevens
Lincoln, Nebraska

DILL PICKLES

Best Ever Dills

Cucumbers
Grape leaves
Dill
5 cups vinegar
3/4 cup salt
7 cups water

Put grape leaves in bottom of jars. Wash and pack cucumbers as tight as possible, using wid-mouth jars if possible. Put generous amount of dill on top. Fill jars with liquid made of vinegar, salt and water. Bring to boil. Seal and set aside for about 6 wk.

Mrs. H. L. Schwenker
Lincoln, Nebraska

B's Dill Pickles

Cucumbers
1 pt. vinegar
½ cup pickling salt
1 cup sugar
Dill
Water

Chunk cukes, pour boiling water over them and let stand 5 min. Put into jars. Mix and heat vinegar, pickling salt, sugar, dill and 1 qt. water and pour over cukes. Seal.

Mrs. Harvey Wagner
Dorchester, Nebraska

Cherry Dill Pickles

Cucumbers
1 tsp. mixed spices
  (Remove red peppers)
1 cup vinegar
1½ cups sugar
Cherry leaves
Dill
Salt

Wash cukes, wipe dry, cut in 1-in. cubes. Put in jar--not too solid. Put spices in bag and boil in vinegar. For 1 qt. use 1 tsp. mixed spices (remove red pepper), 1 cup vinegar and 1½ cups sugar. Let come to boil and seal. Boil spices with water and add liquid to jar. To cure pickles, put layer of cukes, one of cherry leaves, then one of dill. Continue until all cukes are used. Cover with brine consisting of 1 qt. salt to 5 qt. water. Let stand or set 14 days.

Mrs. Dale A. Rogers
Lincoln, Nebraska

Cherry Dills

1 cup pickle salt
5 qt. water
Cherry leaves
Dill
3 cups sugar
1 cup vinegar

(continued on next page)

Dissolve pickle salt in water. Cover with cherry leaves and dill. Mix. Pour over cucumbers (cold solution). Let set 10 days or until cured. Take out of solution, wash, cut into chunks and cover with syrup. Cook syrup (sugar and vinegar) until thick, pour hot on cucumbers and seal.

Mrs. Dick Ruser
Grant, Nebraska

Cold Water Dills

Medium-size cucumbers
Dill
Garlic
Pickling salt
Pickling spices
Cider vinegar

For ½ gal. jar, use 3 ro 4 heads of dill, 2 or 3 cloves of garlic (optional), ½ cup minus 2 tsp. pickling salt. Take 1 cup cider vinegar, fill jar with cold water, seal tight with zinc lids, shake until salt is dissolved. Let stand in warm place 2 or 3 days, then store in cool place. Ready to use in 5 to 6 wk. Can add some pickling spices (1 or 2 tsp.) if desired when mixing.

'Another Backyard Farmer'

Dill Pickles

Cucumbers
6 cups water
½ cup canning salt
2 cups vinegar
½ tsp. powdered alum
Clove
Garlic
Dill

Boil water, salt, vinegar and alum. Wash and pack pickles in jars. Put in clove of garlic and dill. Use hard water instead of city water and pickles will be firm in top mixture.

Mrs. John Dominick
Mission Hill, South Dakota

Dill Pickles

Cucumbers
6 qt. water
1 pt. salt
1 tsp. alum
1 qt. vinegar
Red pepper
Black pepper
Horseradish
Bay leaf
Dill

Wash cukes and dry with cloth. Pack in jars with dill stems and heads. Have jars clean and hot. Boil water, salt, alum and vinegar and pour, boiling hot, over cukes and seal. If you wish a spiced dill you may add to each qt. jar 1 red pepper, 1½ tsp. whole black pepper, horseradish the size of a nickel and 1 small bay leaf.

Wilma Nyffeler
Alliance, Nebraska

Dill Pickles

Cucumbers
5 cups vinegar
7 cups water
3/4 cups salt
Green dill

Put washed cucumbers in glass jars. Add a lot of green dill. Pour over them the boiling vinegar, water and salt. Seal the jar. Pickles will be ready for use in 2 wk.

Mrs. Zelda Stuck
Auburn, Nebraska

Dill Pickles

1 bu. small cucumbers
7 qt. water
1 qt. vinegar
3 cups salt
Dill
Pepper

Put piece of dill and pepper in every jar. Bring water, vinegar and salt to boil. Chop dill and pepper and put in each jar. Pack clean cucumbers in jars and cover with hot liquid. Seal at once. Use in 2 wk.

Mrs. L. Sarson
Bassett, Nebraska

Dill Pickles

Cucumbers
Dill
6 qt. water
1¼ cup salt
1 tsp. alum
4 cups vinegar

Clean cucumbers and pack in fruit jars with dill in top. Boil water, salt and alum. Then add vinegar and boil again. Pour over cucumbers and seal. Store 6 to 8 wk. Old-fashined Mason lids and rubbers are better for sealing pickles.

Mrs. Raymond J. Hillman
Bellevue, Nebraska

Dill Pickles

Dill-size cucumbers
1 ripe dill head
4 tbsp. coarse salt
1½ cups cider vinegar
Pinch of alum
Garlic

Wash cucumbers and pack in ½ gal. jar (or 2 qt. jars). Add dill head, salt, vinegar and alum. With boiling water, finish filling the jar and seal. Turn jar several times to dissolve salt. Ready to use in 2 mo. (Garlic may be added.)

Mrs. Marie Jonas
Clarkson, Nebraska

Dill Pickles

Cucumbers
Dill
Garlic
Cider Vinegar
Pickling salt
Water

Fill jars with medium-size cucumbers. On top of jar, add a large head of dill; also put a head on bottom of jar. If you like garlic, add a small piece on bottom of jar and also on top. Mix 3 cups water, 1 cup cider vinegar and ¼ cup pickling salt, boil and pour hot over cucumbers. Seal. (Be sure boiling water, cider vinegar and pickling salt are well mixed.)

Mrs. Charles Weber
Columbus, Nebraska

Dill Pickles

Cucumbers
Dill
4 cups water
1 cup vinegar
¼ cup pickling salt

Select fresh cucumbers, wash and pack in sterilized jars. Place dill on top of cucumbers. Boil water, vinegar and pickling salt. Cool. Pour over cucumbers and seal. Liquid will become cloudy in a few days but will clear. Pickles will be crisp, firm and evenly colored.

Mrs. Edward Havlat
Crete, Nebraska

Dill Pickles

12 lb. cukes
Dill
1 oz. (4 tbsp.) pickling spices
2 cups vinegar
8½ qt. soft water
2 cups pickling salt
Grape leaves (or Swiss chard)
Garlic

Cover bottom of 4 gal. crock with layer of leaves, layer of dill, ½ of spices. Wash cukes and arrange in crock to 2 or 3 inches from top. Add a layer of dill, remaining spices, layer of leaves. Mix water, salt and vinegar and pour over. Put a weighted plate on top and cover with a clean cloth. Set in a warm place (80° to 85°). Skim off sum every few days. Be careful not to mix it into the brine. After active fermentation ceases, draw off brine and heat to boiling (or make fresh). Pack cukes into sterile glass jars, add garlic and grape leaf if desired, also dill. Pour hot brine over and fill jars. Seal.

Mrs. Robert Reckling
Crete, Nebraska

Dill Pickles

Cucumbers
Dill
2 cups vinegar
1 cup water
2 tbsp. salt
3 tbsp. sugar

(continued on next page)

This recipe is for 2 qt. Boil water, vinegar, salt and sugar and pour over pickles and dill. Seal. These will also keep in open jar.

Dill Pickles

Cucumbers
Dill
2 qt. water
1 qt. vinegar
1 small cup salt

Boil water, vinegar and salt. Pour over pickles in jars, either hot or cold. Put 1 head of dill in each qt. jar. Real good pickles with this recipe.

Mrs. J. B. Teeters
Decatur, Nebraska

Dill Pickles

Cucumbers
1 qt. vinegar
1 cup boiling water
½ cup pickling salt
1 cup sugar

Put dill, a slice of onion and a grape leaf in bottom and on top of each jar. Slice cucumbers ¼ in. thick and fill jars. Cover with the vinegar, water, salt and sugar mixture. Put on cold and seal jars.

Mrs. Harvey Wagner
Dorchester, Nebraska

Dill Pickles

Cucumbers
1 cup vinegar
¼ cup pickling salt
Dill

Boil water, vinegar and salt. Pour hot over cucumbers that have been packed in jars with some dill. Seal.

Mrs. Harvey Wagner
Dorchester, Nebraska

Dill Pickles

Cucumbers
Dill
2 tbsp. salt
Garlic
Pickling spice
Water

Use any size cucumbers. Put dill in bottom of jars; fill up with cucumbers. Put in 2 level tbsp. salt, a small piece of garlic and pickling spice. Put more dill on top, fill jars with cold water and seal.

Mrs. Harvey Wagner
Dorchester, Nebraska

Dill Pickles

Medium-size cucumbers
Dill
1/8 tsp. powdered alum
1 cup vinegar
Garlic cloves
1 tsp. mustard seed
2 cups water
1 tbsp. pickling salt

Scrub medium-size cucumbers with brush and pack them in hot sterilized jars. To each quart add 2 heads of dill, 1 clove fresh garlic, 1/8 tsp. powdered alum and 1 tsp. mustard seed. Mix vinegar, water and salt. Heat to make brine. Fill each quart with hot brine. After filling jars, seal, turn upside down a couple times and let cool before storing. As they cool, they will turn a nice shade of yellow. Keep for 2 wk. before opening.

Melinda Koermer
Elkhorn, Nebraska

Dill Pickles

Cucumbers
Dill
1 qt. vinegar
1 cup pickling salt
3 qt. water

Mix in open jar or crock. Put in enough cucumbers so the brine will cover. Turn a plate over the top and weight it to hold them down. Let stand overnight. Drain and heat the liquid boiling. Pack pickles in glass jars with dill and pour the boiling liquid over them to fill the jar. Seal while hot. Makes about 8 qt. ready to use in 2-3 wk.

Mrs. Cyrus Tharp
Gothenburg, Nebraska

Dill Pickles

Cucumbers
1 pt. vinegar
Dill
Grape leaves
Cherry leaves
1 cup salt
3 qt. water

Bring vinegar to a boil and put mixture hot on pickles.

Mrs. R. F. Evans
Hickman, Nebraska

Dill Pickles

Cucumbers
Dill
Grape leaves
1 pt. vinegar
1 cup salt
2¼ qt. water

Wash cucumbers and pack in jars--leaving room for dill and grape leaves. Bring brine to boil--cover pickles and seal.

Mrs. Laddie Jirovsky
Hooper, Nebraska

Dill Pickles

Cucumbers
Dill
Garlic
5½ cups vinegar
1 tbsp. sugar
Chili pepper
1 cup canning salt
13½ cups soft water

Boil mixture of water, vinegar and canning salt 15 min. Pour hot solution over pickles carefully packed in glass jars. To each jar add 1 tbsp. sugar, a clove or two of garlic, a piece of chili pepper and plenty of dill. Seal tightly, put in a container of boiling water and leave them there until water is cold.

Esther Bruneau
Jefferson, South Dakota

Dill Pickles

Cucumbers
Dill
Garlic
3 cups vinegar
1 cup salt
9 cups water

Wash but don't cut pickles. Pack into jars, put dill in bottom and then the pickles. Mix salt, vinegar and water. Bring to a boil. Pour over the pickles, put dill on top and seal. Put a little garlic in each jar. The zinc lids may bulge or fizz some as if pickles are spoiling but don't open. Just let them be and they will be good after they are through working.

Noelle Bros
Laurel, Nebraska

Dill Pickles

4 lb. 4-in. cucumbers
Dill (2 or 3 heads)
3 cups white vinegar
6 tbsp. salt
3 cups water

Wash cucumbers. Cut in half lengthwise. Combine salt, vinegar, water. Heat to boiling. Pack cucumbers into clean hot jars. Add dill to each jar. Fill with boiling pickling syrup to within ½ in. of jar top. Adjust lids. Process in boiling water 10 min. Makes 7 pt.

Mrs. W. L. Kinghorn
Lewiston, Nebraska

Dill Pickles

Cucumbers
Pinch alum
Dill
Grape leaf
½ onion per quart
3 cups vinegar
1 cup canning salt
9 cups water

Bring mixture of salt, vinegar, water and alum to boil. Pour over cucumbers. Put in bottom of jar a grape leaf, dill and ½ onion. Seal. Ready in 3 wk.

Nellie Dwinell
Lincoln, Nebraska

Dill Pickles

20-25 4-in. cucumbers
2 whole allspice
1/8 tsp. powdered alum
5 cassia buds
2 whole cloves
2 heads green dill
1 clove garlic, cut
6 whole pepper berries
1 tsp. white mustard seed

Wash cukes; let stand in cold water overnight. Pack into hot, sterilized quart glass jars. Add ingredients to each quart. Combine 1 qt. vinegar, 1 cup salt, 3 qts. water. Heat to boiling and fill the jars. If available, place a grape leaf in each jar and seal. Makes 6 to 8 qt.

M. Folsom
Lincoln, Nebraska

Dill Pickles

Cucumbers
Dill (stalk)
Grape leaf
1 qt. vinegar
1 cup salt
2 qt. water

Wash pickles. Put a grape leaf in bottom of each jar. Heat water, vinegar, salt and pour over pickles. Put a stalk of dill on top of each jar before sealing. I have used 1 tsp. dill seed to each qt. when fresh dill is not available.

Mrs. A. J. Hindmand
Lincoln, Nebraska

Dill Pickles

Medium-size cucumbers
Cherry leaves
Dill
Grape leaves
1 cup vinegar
½ cup coarse salt
1 gal. water

Take medium-size cucumbers. Let stand in water overnight or 24 hr. Remove, wash and dry. Put grape leaves on bottom of stone jar, then pack in layer of cucumbers very solid, cover with dill and cherry leaves. Repeat this process, being sure cucumbers are packed solidly. Use plenty of dill. When jar is full, cover with dill, cherry leaves and grape leaves. Some horseradish may be added. For a 2 gal. jar, mix salt, vinegar and water until salt is dissolved. Pour over cucumbers and place a weighted plate on top. It is essential that cucumbers are held down. Put into a cool cellar 10 to 12 days. Remove pickles as needed, replacing weight each time.

Mrs. S. C. Huffman
Lincoln, Nebraska

Dill Pickles

Cucumbers
Cherry leaves
Dill
Garlic
Grape leaves
1 red pepper
Salt
Water

(continued on next page)

Use stone jar. To make 2 gal., the first layer is dill, garlic, grape and cherry leaves and 1 red pepper. Then put a layer of cucumbers about 5 in. long and repeat until jar is filled. Cover with boiling water in which has been dissolved a little more than two handsful of salt. Cover with plate and stone. If more pickles are added, pour lukewarm water on. Let stand 1 wk. to sour, then put in cellar.

Marie Kuska
Lincoln, Nebraska

Dill Pickles

Cucumbers
Dill
Garlic
1 cup vinegar
½ cup salt
9 cups water

Put a layer of dill in bottom of jars. Also clove of garlic. Pack jar full of fresh, clean cukes. Set jars in canner rack. Bring to boil water, vinegar and salt. Pour solution over pickles while boiling hot. Quickly place lids on and place in canner of boiling water to cover jars and bring to rolling boil. Lift rack out of canner. Ready in 6 wk.

Mrs. John Lustrea
Lincoln, Nebraska

Dill Pickles

Cucumbers (fresh)
Dill
1 cup vinegar
¼ cup salt
1 qt. water

Use fresh cucumbers from garden. Put them in jars and cover with boiling syrup of salt, vinegar and water. Add as much fresh dill--seeds, stems--as you can and seal hot. Don't put them away in dark for at least 1 wk.

Mrs. Clarence Meyer
Lincoln, Nebraska

Dill Pickles

Cucumbers
Dill
Garlic
Red pepper (pinch)
½ cup vinegar
1 tbsp. salt
Water

Soak cucumbers in cold water overnight. Wash and put in jars with dill and garlic. Add a pinch of red pepper. Add 1 tbsp. salt to 1 qt. water. Put jars into hot water bath and cook 5 min. Remove and add ½ cup hot vinegar and fill rest of jar with boiling water. Screw shut.

Mrs. L. C. Rieger
Lincoln, Nebraska

Dill Pickles

Cucumbers
Dill
1 cup pickling salt
1 qt. vinegar
3 qts. water

Wash pickles. Put in jar with dill on top and around pickles. Boil vinegar, water and salt. Pour over pickles while hot. Seal.

Mrs. C. M. Stevens
Lincoln, Nebraska

Dill Pickles

Cucumbers, 1 pk. medium size
Dill
Garlic
1 qt. vinegar
1 cup pickling salt
3 qt. water

Combine vinegar, water and salt. Bring to boil. Wash pickles. Pack in glass jars, cover with boiling liquid. Add 2 hearts of dill and 1 clove garlic and seal.

Mrs. Richard Walin
Lincoln, Nebraska

Dill Pickles

35-40 fresh cucumbers
2 tbsp. mixed spices
½ lb. dill
2 cups salt
2 gal. water
2 cups vinegar

Wash and dry cucumbers. Put a layer of dill and ½ of the spices in a stone jar. Add cucumbers. Put the remaining spices and dill on top of cucumbers. Boil salt, water and vinegar 2 min. Cool to room temperature and pour over cucumbers. Cover with a plate weighted down to hold the cucumbers under brine. Keep at an even temperature (75°-85° F). Remove scum each day. Pickles are ready for canning when they are crisp, uniform in color and well flavored with dill. This usually requires 2 to 4 wk. Pack cured pickles in hot jars, cover with hot brine; seal at once. If to be stored a long time, dill pickles should be processed in water bath 15 min. at simmering.

Mrs. Glen Miller
Louisville, Nebraska

Dill Pickles

Cucumbers
1/8 tsp. powdered alum
Dill
Red pepper
Garlic
1 qt. vinegar
1 cup salt
3 qt. water

(continued on next page)

Dill Pickles

Cucumbers
1/8 tsp. powdered alum
Dill
Red pepper
Garlic
1 qt. vinegar
1 cup salt
3 qt. water

Let fresh cucumbers stand overnight in cold water. Pack in clean, sterile jars. Put 1/8 tsp. powdered alum, 2 heads of dill, 1 small hot red pepper and 1 clove of garlic in each quart jar. Combine vinegar, salt and water. Heat to boiling. Fill jars with liquid and seal.

Lil
Minden, Nebraska

Dill Pickles

Cucumbers
Dill
1 cup cider
¼ cup canning salt
1 qt. water

Heat water, vinegar and salt. Pack dill-size cucumbers lengthwise in jars. Cut them in fourths lengthwise. Put in as much dill as desired. Pour the hot liquid into the jars. Seal. Place jars in an enamel canner or large container with a trivet of some sort under them. Add enough hot water to come up the sides. Bring to a boil. Turn off heat and let jars set in the water until it cools.

Mrs. Richard T. Kucera
Mitchell, Nebraska

Dill Pickles

Cucumbers
Dill
1 qt. vinegar
1 cup salt
3 qts. water

Heat to boiling, pour hot over cucumbers packed in a jar, with a sprig of dill on the top.

Mrs. Ora Randall
Mitchell, Nebraska

Dill Pickles

Cucumbers
Dill
1 cup vinegar
3 tbsp. salt
3 cups water

Put pickles in and simmer, then can as usual with dill.

Mrs. Geo. Schmid
Monroe, Nebraska

Dill Pickles

Cucumbers
Alum
Dill
Grape leaves

½ cup vinegar
½ cup salt
13 cups water

Boil and cool water, vinegar, salt and pinch alum. Pour over pickles. Put dill and grape leaves on bottom and top of jar.

Mrs. Alvin Dvorak
Morse Bluff, Nebraska

Dill Pickles

Cucumbers
1 tsp. alum
Garlic beans
Dill

1 qt. vinegar
1 pt. salt
6 qt. water

Boil water, vinegar, salt and alum. Pack whole cucumbers in jars with dill. Cover with hot liquid and seal. If you like garlic, add several garlic beans.

Dill Pickles

Medium-size cucumbers
Dill
1 cup grated horseradish
1 cup mustard seed

Red pepper
1 qt. vinegar
3½ cups pickling salt
14 qt. water

Soak medium-size cucumbers overnight in brine made of 1½ cups pickling salt to 4 qt. water. Make a canning brine of 10 qt. water, 1 qt. vinegar, 2 cups salt. Boil 10 min. Let this brine stand overnight. In morning, drain cucumbers and place in sterilized jars with small bunches of dill. If desired, 1 cup grated horseradish and 1 cup of mustard seed may be used with dill by placing a small amount of each in each jar. Pack cucumbers tightly. Add small red pepper to each jar if desired. Cover cucumbers in jars with cold canning brine, being sure brine covers cucumbers. Put on cap, screwing band tight. These will ferment 3 or 4 days.

Mrs. James Rader
Nebraska City, Nebraska

Dill Pickles

Cucumbers
Dill
1 cup vinegar

½ cup salt
4 cups water
Grape leaves

Bring vinegar, salt and water to boil. Pour over pickles while hot. Add grape leaves and dill.

Esther Langhorst
Nickerson, Nebraska

Dill Pickles

Cucumbers
Alum
Dill
Carrot
Horseradish

Onion
½ cup vinegar
½ cup salt
4 cups water

Boil brine of water, vinegar and salt 5 min. Pour over pickles in jar and seal. Along with cucumbers, place 2 small pieces of horseradish, carrot and small onion in each jar. Place dill and alum size of hickory nut in bottom of jar. May use rhubarb in place of horseradish.

E. H. Williams
Norfolk, Nebraska

Dill Pickles

Cucumbers
Dill
Garlic

1 qt. vinegar
1 cup salt
3 qt. water

Heat salt, vinegar and water to boiling. Pour over pickles, washed and packed, with dill in qt. jars. Makes 6 qt. pickles. Clove of garlic in bottom of jar is good also. Seal and store. Let stand to cure about 3 or 4 wk. before using.

Mrs. J. L. Scroggin
Oak, Nebraska

Dill Pickles

Cucumbers
Dill
3 cups vinegar

1 cup salt
9 cups water

Put pickles in 2 quart cans with a stalk of dill. Bring salt, water and vinegar to boil and allow to cool. Fill jars and seal.

Mildred Saf
Oakland, Nebraska

Dill Pickles

Cucumbers
Dill
Garlic

1 qt. vinegar
1 cup salt
3 qt. water

Wash 5-6 in. cucumbers in cool water and pack into jars with 1 clove of garlic and 3 sprigs of dill to each quart. Boil vinegar, salt and water for 5 min. Cool slightly. Pour over cucumbers and seal.

Mrs. Maurice B. Allstot
Omaha, Nebraska

Dill Pickles (Crock Style)

Cucumbers
Dill
Bread

Salt
Water

Wash medium-size cucumbers and slit each lengthwise into four sections. Don't slice all the way through--leave them whole. Into a 2 or 3 gal. crock put about 3 branches of washed dill and a handful of coarse salt. Fill the crock with cukes. Put about 3 more branches of dill on top and some more coarse salt (about 3 tbsp.). Put 1 piece of bread, preferably the heel, on top. Homemade bread is best. Fill the crock with lukewarm water about 2 in. above the cucumbers. Put a plate on top of cucumbers to weight them down. Let these stand in a warm room about 6 days. After 4th day take bread out. Pickles are ready to eat in about 6 days. You can pack pickles in jars, pour liquid over them and store in refrigerator, or you can can pickles by putting into jars, heating liquid brine to boiling, pouring over pickles and sealing. These will keep about 6 mo.

Mrs. D. W. Moench
Omaha, Nebraska

Dill Pickles

Cucumbers
Dill
1 cup salt

1 qt. vinegar
1 tbsp. mixed pickling spice

Boil 5 min. Pour over pickles packed in sterilized jars with dill. Seal while hot.

Mrs. Elmer Pfeiffer
Oxford, Nebraska

Dill Pickles

Cucumbers
Dill
Alum
Grape or horseradish leaves

1 qt. water
1 pt. vinegar
½ cup sugar
½ cup salt

Cook water, vinegar, salt and sugar until it boils. Pour boiling hot water on pickles, let stand 8-10 min. Pour off hot water. Put dill on bottom and top of cucumbers. Put a little alum on top. Can also put grape leaves or horseradish leaves on top. Pour hot liquid over pickles in jars and seal.

Rose A. Houska
Seward, Nebraska

Dill Pickles

Cucumbers
Dill
Alum
Sugar

13 cups water
6 cups vinegar
1 cup salt

(continued on next page)

Wash cucumbers and prepare as desired--slices, chunks or sticks. Combine water, vinegar and salt. Boil. Pack pickles in jars. Add head of dill to each jar. To each jar add 1 tsp. sugar and dash of alum. Pour hot vinegar mixture over pickles. Seal. Now wrap your jars in a blanket until they are cold. These are ready to eat almost immediately.

Geraldine Walker
Sidney, Nebraska

Dill Pickles

Cucumbers
Dill
Onion (if desired)
2 cups water
1 cup vinegar
2 tbsp. salt
Alum size of pea

Combine water, vinegar, salt and alum and bring to boil. Fill jars with pickles adding dill (onion). Pour vinegar solution in jars of pickles and seal. Put in canner of boiling water. Let stand until cold and remove.

Mrs. Richard Evers
Syracuse, Nebraska

Dill Pickles

Cucumbers
Dill
Garlic
1 pt. vinegar (white or brown)
1 cup pickling salt
2¼ qt. water
1 tsp. alum

Put plenty of dill in sterilized jars. Add 1 clove garlic to qt. and 2 or 3 to ½ gal. jars. Pack cucumbers in jars. Put vinegar, salt, water and alum in large kettle and bring to boil. Let boil until salt and alum are dissolved. Pour over cucumbers. Keep solution boiling hot as it is poured into jars. Use rubbers and zinc lids on jars.

Mrs. Howard Huddle
Valentine, Nebraska

Dill Pickles

Cucumbers
4 cups water
1 cup vinegar
½ cup pickling salt
Dill

Bring water, vinegar, salt to boil. Boil 5 min. Pour over washed pickles packed in jars with dill. Seal. This liquid may be used hot or cold.

Mrs. Nettie Lovejoy
Valentine, Nebraska

1-3-9 Dill Pickles

Cucumbers
1 cup salt
3 cups vinegar
9 cups water
¼ tbsp. dill

Mix salt, vinegar, water and dill until salt is dissolved. Pour cold over pickles and seal.

Mrs. Ed Scheuneman
Ithaca, Nebraska

Dill Pickles

Cucumbers
4 cups water
1 cup vinegar
Dill
½ cup bulk salt
Grape leaves
Garlic or onion (if desired)

Bring water, vinegar and salt to boil. Wash and dry cucumbers and few grape leaves. Put grape leaf in bottom of jar (1 or 2 qt. jars) and fill with cucumbers, putting in sprig of dill. A few slices of garlic or onion may be added. Put grape leaf on top. Fill with boiling liquid. Seal.

Mrs. Marion Spangler
Weeping Water, Nebraska

Dill Pickles

Cucumbers
9 cups water
2 cups vinegar
½ cup salt
Dill
Grape leaves

Bring water, vinegar and salt to boil, pour over pickles packed in stone jar with dill. Grape leaves can be placed at bottom of jar and also on top. Cover jar with a clean cloth.

Bessie Meduna
Weston, Nebraska

Dill Pickles

Cucumbers
Grape leaves
Dill
6 qt. water
1 lb. coarse salt
1 tsp. alum
4 cups vinegar

Wash pickles, put in Mason jars with grape leaves and dill in top and bottom of jar. Boil water, salt, alum and vinegar. Pour over pickles hot and seal. This recipe makes about 15 qt.

Miss Melva Risse
West Point, Nebraska

Dill Pickles

Cucumbers
13 cups water
6 cups vinegar
1 cup salt
Dill
2-3 diced garlic cloves
1 cup sugar

Boil water, vinegar, salt. Pack cucumbers in jars with dill and diced garlic cloves. Sprinkle sugar on top and pour boiling mixture over cucumbers. Seal and place in hot water. Let cool naturally.

Mrs. Charles Bors
Wilber, Nebraska

Dill Pickles

35-40 cucumbers
3 tbsp. mixed spices
Garlic
Bay leaf
½ tsp. mustard seed
Dill
1½ cups salt (non-iodized)
2 cups vinegar
2 gal. water
Red Pepper

Wash and dry cucumbers. Put a layer of dill and ½ of the spices in a stone jar. Add cucumbers. Add salt and vinegar to 2 gal. hot water. Stir until salt dissolves. Cool brine. Pour over cucumbers. Cover with dinner plate. Fill jar with water and use to hold plate below brine. Keep at even temperature 68-72 degrees 2 to 4 wk. Remove scum each day. Drain brine from cucumbers and boil 5 min. Pack cucumbers in jar and add 1 clove garlic, 1 bay leaf, ½ tsp. mustard seed and 1 piece of red pepper. Cover with brine, seal according to type of jar lid used and process pints and quarts 10 min. in a boiling water bath.

Mrs. Edson
Council Bluffs, Iowa

Dill Pickles

Cucumbers
17 cups water
1 cup salt
½ cup vinegar
Strips of green pepper
1 clove garlic
1/8 tsp. alum
Dill

Boil water, salt and vinegar. Pour over pickles and seal.

A. L. Wallenberg
Sioux Falls, So. Dakota

Dill Pickles

40-50 cucumbers
10-15 grape leaves
1 pt. cider vinegar
1 lb. coarse salt (Kosher or pickling)
Dill
¼ cup sugar (if desired)
2 tsp. alum
2 gal. water
2 tbsp. pickling spice (in bag)

(continued on next page)

Wash uniform-size cucumber, not too large or ripe. Put 10-15 grape leaves in bottom of crock. Alternate layers of cucumbers with plenty of dill. Heat and cool vinegar, salt, sugar, water, alum and pickling spice. Layer of dill on top. Place brine over cucumbers with plate on top. Weigh down so cucumbers are under brine. Cover with cloth-skim scum each day. Store in cool place-68-72 degrees for 2-4 wks. If necessary, during the curing process make additional brine and pour over cucumbers to be sure they are completely covered at all times. Pickles are cured when well-flavored and even in color. There should be no white spots. Remove pickles from crock--place in jars. May slice if necessary. Leave ¼ in. head space. Prepare fresh brine, hot (3/4 cup salt 1 cup vinegar, 4 cups water) boil 5 mins. or strain brine from crock and boil 5 min. Pour over cucumbers--leave ¼ in. head space. Pour over pickles. Adjust caps, process pints and quarts 15 min. in boiling water bath or 5 min. at 15 lb. pressure in pressure cooker. Allow to cool naturally before opening cooker. Add small amount of garlic for kosher style.

Mrs. Lawrence E. Kokulil
Omaha, Nebraska

Dill Pickles in Jar or Barrel

Cucumbers
1 cup salt
1 gal. water
Cherry leaves
4 garlic bunches (to 10 gal.)
1 tbsp. alum (to 10 gal.)
Dill
Horseradish

One cup salt to 1 gal. water. 4 garlic bunches to 10 gal. 1 tbsp. alum to 10 gal. Dill, cherry leaves and horseradish.

Mabel Coleman
Merna, Nebraska

Dill Pickles--Long Method

35-40 cucumbers
Dill
2 cups vinegar
2 gal. water
3 tbsp. mixed pickling spice
1½ cups salt (pickling)
Grape leaves

Wash and dry cucumbers. Place a layer of dill in stone jar. Add cucumbers. Put layer of dill and rest of spice on top. Add salt and vinegar to 2 gal. hot water. Stir until salt is dissolved. Cool brine and pour over cucumbers. Cover with a plate and weight down. Keep at an even temperature (68-72 degrees) 2 to 4 wks. Remove scum each day. The pickles are ready for canning when well flavored and of even color throughout. There should be no white spots. Pack the pickles into hot jars. Make new brine or strain brine in which cured. Boil 5 min. Pour over pickles. Process pints and quarts 10 min. in boiling water bath. For kosher type pickles add 1 clove garlic, 1 bay leaf, ½ tsp. mustard seed, 1 piece red pepper to each jar when canning. You can use a layer of washed grape leaves in jar when curing if desired.

Mrs. Thomas McLellar
Spencer, Iowa

Dill Pickles Made With Cold Water

Cucumbers
Dill
Garlic
Grape leaves

5 tbsp. salt
½ cup vinegar
5-6 cups cold water

Pack pickles in jars with dill and garlic if you like. You can use grape leaves too. Pour over this salt, vinegar, water. Mix before pouring. Makes 2½ gal. pickles.

Mrs. George Schmid
Monroe, Nebraska

Dill Pickles (No Vinegar)

Cucumbers
2 tbsp. salt
1 qt. water

Dill
Garlic (if desired)

Wash cucumbers in warm water, remove any imperfections. Pack in an open jar so they can be weighted down to keep under the brine. Use a 1 gal. stone jar. The brine will develop a little scum. Remove daily but keep at room temperature so they will ferment. These will be ready in about 10 days. After pickles are cured, put in quart jars and pour some of the brine they were in over them and put in refrigerator. Make brine of 2 tbsp. salt to 1 qt. water. Use warm water to make sure salt is dissolved. When cool pour on cucumbers. Use fresh dill stems and heads. Pack with cucumbers in jar. Garlic may be used if desired. These are nice for anyone that cannot have vinegar.

Wilma Nyffeler
Alliance, Nebraska

Dill Pickles--Open Jar

Cucumbers
3 gal. water
2 lb. coarse pickling salt

Grape leaves
Dill (fresh on stalk)
Cabbage leaves

Boil and skim 3 gal. water plus 2 lb. coarse pickling salt. Select cucumbers 6-8 in. Wash and dry. Put a layer in stone jar or crock that will hold 4 gal. Put in layer of grape leaves, layer of dill (fresh on stalk), proceed until jar is filled. Lay cabbage leaves on top to collect skum. Put a large weight (jar of water) fill with the brine, let stand 2 or 3 wks. Skim off as scum collects. A cloth can be used instead of cabbage leaves. Wash off as scum collects. A cloth can be used instead of cabbage leaves. Wash off as scum forms.

Mrs. J. L. Scroggin
Oak, Nebraska

Dill Pickles--Open Jar

Cucumbers
Dill
Cherry leaves
Grape leaves
Salt
Vinegar

Use a stone jar. Line bottom with dill, cherry leaves and grape leaves. Sprinkle salt over this. Place layer of cucumbers side by side. Do not pile. Repeat layers until desired amount, ending with leaves. Cover with water and a little vinegar. Place weight over this. Tie a cloth over jar and place in not too cool spot to cure. Sometimes I slice dill pickles into a fruit jar and cover with vinegar and light brown sugar solution. Sometimes I add a little water so they won't be so sharp.

Fannie Janike
Rising City, Nebraska

Dill Pickles--Qt. Jar Method

Cucumbers
1/8 tsp. alum
1 clove garlic
2 heads dill
1 bay leaf
2 cups cider vinegar
6 cups water
½ cup coarse salt

Wash 24 large cucumbers. Cover with cold water, let stand overnight in cool place. Pack in 3 qt. jars, leave room for spices. Add to each jar 1/8 tsp. alum, 1 clove garlic, 2 heads dill, 1 bay leaf. Combine and bring to boil the vinegar, water and salt. Pour over and seal.

Mrs. Jay Collins
Alliance, Nebraska

Dill Pickles--Sweet

Cucumbers
1 cup salt
1 gal. water
Dill
3 cups vinegar
3 cups sugar
2 cups water
Whole mustard seed

Clean and put cucumbers in cold salty water until they turn yellow (1 cup salt to 1 gal. water). This may take a few days or up to 1 wk. Make a solution of vinegar, sugar, water and whole mustard seed. When it starts to boil, add cucumbers which have now been cut into halves or quarters, depending on size. Heat until they are heated through. Place in jars with plenty of dill.

Mrs. Ed Scheuneman
Ithaca, Nebraska

Dills

Cucumbers
Dill
Clove garlic
1 cup salt (coarse)
3 cups vinegar
9 cups water

(Continued on next page)

Wash dill-size cucumbers and pack in clean jar with 2 or 3 heads of dill. Prepare liquid of salt, vinegar and water. Bring to boil, let cool and pour off cucumbers. Seal. For a different flavor add a clove of garlic.

Dorothy Straw
Lincoln, Nebraska

Dills

Cucumbers
Cherry leaves
Grape leaves
Dill
1 qt. vinegar
1 gal. water
2 cups salt

Put 8-12 cherry or 2 grape leaves in bottom of qt. jar. Pack in cucumbers with 2 medium(I use large) sprigs of dill (leaves, stems and seeds) about ½ way up in jar. Cover with boiling liquid of vinegar, water and salt. Seal. Leave in warm place until after fermentation. Less sour and salty and very dilly.

Mrs. Jessie L. Salfrank
Rock Port, Missouri

Dills in Open Jar

Cucumbers
Grape leaves
Dill
2 lb. coarse salt
3 gal. water
Cabbage leaves
1 qt. vinegar
3½ lb. sugar
1 tbsp. mixed spices

Put dill-size cucumbers in layers putting grape leaves and dill between layers. Add 2 lb. coarse salt to 3 gal. water--boil--pour over--put layer cabbage leaves over top. Use clean cloth and weight. Takes 2 to 3 wks. Wash cloth frequently. To make sweet pickles of these soak dills 3 days in cold water, cut in half. First day add 1 tbsp. alum to 4 qt. water, boil, pour over pickles, leave 24 hrs. Second day, wash, put in 3 qt. water and 1 tsp. ginger. Simmer 30 min. Pack pickles in syrup of vinegar, sugar and mixed spices. Pour this over cucumbers 3 mornings. Do not seal.

Mrs. Nellie Dwinell
Lincoln, Nebraska

Dutch Dill Pickles

Cucumbers
1 sliced onion
1 stalk celery
Garlic clove
Pickling spice, if preferred
3 cups vinegar
1 cup water
¼ cup salt
½ cup sugar
¼ tsp.. alum
Grape leaf
Horseradish

(Continued on next page)

Wash 30 to 35 dill-size cucumbers and pack into jar with onion, celery, garlic clove and pickling spice if used. Make syrup from vinegar, water, salt and sugar. Heat, let cool and pour on jarred cucumbers. Add alum, grape leaf and horseradish to each jar.

Mrs. H. J. Newman
Alliance, Nebraska

Dill Sticks

Cucumbers
3 tbsp. pickling spice
Dill
1 3/4 cup sugar
½ cup salt
4 cup vinegar
4 cup water

Wash and dry 35-40 (4-5 in.) cucumbers, then slit in ¼'s longways. Combine spices (in a bag), sugar, salt, vinegar and water. Bring to boil and simmer 15 min. Place 1 head of washed dill in each jar. Pack cucumbers in jars. If you put in quart jars, place a second head of dill in jar half way up. Pour boiling brine over cucumbers in jar to ½ in. from top. Put lids on and process 15 min. in boiling water bath. Make plenty of these crisp, slightly sweet dills. These always keep. No milky residue in any of the jars.

Mrs. Thomas L. McLellar
Spencer, Iowa

Fermented Dill Pickles

Cucumbers
3/4 cup pickling spices
Dill
2 cups vinegar
1½ cup salt
2 gal. water

Cover 20 lbs. (3½ to 5½ in) cucumbers with cold water. Wash thoroughly, using a vegetable brush handling gently to avoid bruising. Take care to remove any blossoms. Drain on rack or wipe dry. Place half the pickle spices and layer of dill in 5-gal. crock or stone jar. Fill with cucumbers to 3 or 4 in. from top of crock. Mix vinegar, salt and water and pour over cucumbers. (Garlic may be added, if desired) Cover with a heavy china or glass plate or lid that fits inside the crock and use a weight, such as a glass jar filled with water, on top of cover to keep cucumbers under brine. Cover loosely with clean cloth. Keep pickles at room temp. and remove film daily when formed. Film may start forming in 3 to 5 days. Do not stir pickles around in jar but be sure they are completely covered with brine. If necessary, make additional brine, using original proportions. In about 3 wks. the cucumbers become olive-green and any white spots inside the fermented cucumbers will be eliminated by processing. Pack the pickles, along with some of the dill, into clean, hot jars. Add garlic if desired. Cover with boiling brine to ½ in. of top of jar. Adjust lids. Put jars into canner containing boiling water; be sure the water comes over jar tops. Cover canner tightly and start to count processing time. Continue heating and process 15 min. Remove jars from canner immediately. Set jars on wire rack several inches apart to cool. Complete seals unless closures are self-sealing type. Makes 9 to 10 qt.
NOTE: The original brine is usually cloudy as a result of yeast development during fermentation period. If this cloudiness is objectionable, fresh brine may be used to cover the pickles when packing them. Make it with the same proportion of vinegar, salt and water as in the original brine. The fermentation brine is generally preferred for added flavor; it should be strained and heated to boiling.

Mrs. Clara Loberg
Norfolk, Nebraska

Good Dill Pickles

Cucumbers
Dill
2 cups water
2 tbsp. pickling salt
½ cup vinegar
Alum

Wash cucumbers and put in jar. Add fresh dill or 1 tsp. dill seed to each quart jar. Bring water, salt, vinegar and alum to a boil. Pour over cucumbers. Partly seal and let stand to work.

Mrs. A. E. Heins
Columbus, Nebraska

Good Dill pickles

Cucumbers
1 qt. white vinegar
2 qt. water
2/3 cup salt
Alum
Horseradish leaves
Grape leaves

Boil vinegar, water, and salt in covered pan. Wash cucumbers and pack loosely, putting dill on bottom and top of cucumbers. Put ¼ tsp. alum on top of each jar. Pour hot vinegar mixture or brine over pickles. Grape leaves or horseradish leaves can be put on top. Seal. These are good in 6 wks. or later. Makes 5 to 6 qts.

Rose A. Houska
Seward, Nebraska

Easy Dill Pickles

Cucumbers
1 qt. vinegar
3 qt. water
1 cup pickling salt
Dill

Soak cucumbers in clear water overnight. Pack in jars. They can be left whole or sliced. Boil vinegar, water and salt; pour over cucumbers in jar. Add 1 large stalk of dill to each quart. Seal jars. This amount of brine will cover 7½ qts. of cucumbers. Let stand 3 wks. before using.

Mrs. C. A. Valder
Tekamah, Nebraska

Hamburger Dill Pickles

Cucumbers
1 clove garlic
Dill
2½ cup sugar
1 qt. vinegar
1 pt. water
½ cup sugar

Put onion, garlic and dill in jar. Fill with sliced (paper thin) unpeeled cucumbers. Boil sugar, vinegar, water and salt together 2 min. Pour over pickles and seal. Good to use in 6 wks.

Mrs. Wm. J. Murphy
O'Neill, Nebraska

Hamburger Dills

Cucumbers
Onion
5 cups vinegar
2 cups water
4 cups sugar
½ cup pickling salt
1 tbsp. alum
1 tsp. celery seed

Slice cucumbers and onion in jar. Boil vinegar, water, sugar, salt, alum and celery seed and pour over cucumbers in jars and seal. Makes enough syrup for 6 qt.

Lydia Schweppe
Murdock, Nebraska

Hamburger Dills

Cucumbers
Dill
Garlic
2½ cups sugar
1 qt. white vinegar
1 pt. water
½ cup pickling salt

Slice medium-size pickles rather thin, pack in quart jars. Add 2 tsp. dill seed or fresh dill sprigs and a bud of garlic to each quart jar. Boil sugar, vinegar, water and salt and store 3 wks. This amount of liquid should cover 4 qt.

Marge Alfson
Wisner, Nebraska

Hamburger Dills

Cucumbers
Onion
Pepper
1 qt. vinegar
1 cup water
2 cups sugar
½ cup pickling salt

Slice medium cucumbers, onion and red, green or pimiento pepper (optional) into jar. Boil vinegar, water, sugar and salt. Pour hot over pickles and seal jars.

Hans Anderson
Wynot, Nebraska

Hamburger Pickles

Cucumbers
Onion
Garlic
Dill
1 pint water
1 qt. vinegar
2½ cups sugar
½ cup coarse salt
1 tsp. alum

Put a slice of onion, clove garlic and dill in bottom of jars. Slice cucumbers (½ in. thick) and fill jar. Boil water, vinegar, sugar, salt, and alum. Pour over cucumbers and seal jars. Liquid makes enough for 3 qt.

Rose A. Houska
Seward, Nebraska

Honey Dills

Cucumbers	1½ cups honey
Garlic	¼ cup salt
Dill	3 cup vinegar
Alum	2 cup water

Soak cucumbers in ice water overnight. Slice lengthwise and pack in jars. Boil honey, salt, vinegar and water. Pour over cucumbers. Add garlic, dill or red pepper in jar, also 1 large lump of alum. Makes 3 qt.

Blanche Norris
Brewster, Nebraska

Kosher Dill Pickles

Cucumbers	3 qt. water
Dill	1 qt. vinegar
Garlic	3/4 cup canning salt
Alum	

Wash cucumbers, don't use large ones, that have a lot of seeds, they will get soft. Put "fresh" dill, one bunch in each quart jar and one garlic bud. Heat water, vinegar, salt and pinch of powered alum to hard boil and pour over packed jars of pickles. Seal at once. Let stand 3 mo. before using, this lets pickles cure and get good taste of dill and garlic. May omit garlic if desired. Makes 5 qt.

Evelyn Schreiner
Fremont, Nebraska

Kosher Dills

Cucumbers	3 qt. water
Dill	1 qt. vinegar
Garlic	1 cup salt
Red Pepper	½ tsp. powdered alum

Soak 1 peck of dill-size cucumbers overnight in water in refrigerator. Place dill in bottom of jar - fill with cucumbers, put dill on top, also 1 button garlic and 1 red pepper. Fill jars with boiling syrup of water, vinegar, salt and alum. Seal and let stand several weeks.

J. O. Olson
Lincoln, Nebraska

Kosher Dill Pickles

Cucumbers	1 qt. vinegar
1/8 tsp. powdered alum	1 cup salt
1 clove garlic	3 qt. water
2 heads dill	Grape leaves
1 hot red pepper	

(Continued on next page)

Wash 20-25 cucumbers. Let stand in cold water overnight. Pack into hot sterile jars. To each quart add 1/8 tsp. alum, 1 clove garlic, 2 heads dill and 1 hot red pepper. Combine vinegar, salt and water, heat to boiling; fill jars. Place grape leaf in each jar and seal. Makes 6-8 qt.

Robert M. Johnston
Emmet, Nebraska

Kosher Dill Pickles

Cucumbers	2 qt. water
Grape leaf	1 pt. vinegar
Dill	2/3 cup meat salt
Clove of garlic	

Wash cucumbers, pack in jars with grape leaf, dill and clove of garlic peeled. Heat water, vinegar and meat salt. Boil this brine and pour over cucumbers. Seal. Can be eaten in 3 days, better if kept at least 10 days.

Mrs. Jessie Krofta
Table Rock, Nebraska

Garlic Dill Pickles

Cucumbers	Red pepper
Dill	6 qt. water
Garlic	1 qt. cider vinegar
Mustard seeds	1 cup canning salt.

Pack pickles in glass jars with plenty of dill and 2 cloves of garlic, 1 tbsp. mustard seed and 1 small red pepper to each quart jar. Boil water, vinegar and salt at least 5 min. Then pour immediately over pickles. Seal tightly.

Esther Bruneau
Jefferson, So. Dakota

Cold Pack Garlic Dills

Cucumbers	½ cup vinegar
Garlic buds	2 tbsp. sugar
Dill	1 tbsp. pickling salt

Wash fairly large cucumbers and pack into jar in which you have put some dill and garlic buds. Mix vinegar, sugar and salt; pour this over cucumbers and fill jar with cold water, seal. Set jar into cooker or can of water to cover to neck of jar. Just let water come to boil and then remove jars. Let stand 4-6 wks. before using, then cut into round slices.

Mrs. Ivan Hiatt
Bristow, Nebraska

Kosher Dill Pickles

Cucumbers
1 cup salt
8 cups water
2 cups white vinegar
3 cups water
1 tbsp. pickling spices
Dill
1 clove garlic
1 hot red pepper

Soak medium-size cucumbers 24 hr. in brine. Remove from brine and wipe dry. Make a solution of vinegar and water and add pickling spice and several bunches of dill or as much as your taste requires. Let solution come to a boil, add cucumbers and take from fire. Into each jar put 1 clove garlic, 1 small hot red pepper. Pack cucumbers in jars, put a bunch of dill in each jar. Put vinegar back on fire and bring to boil. Pour over cucumbers and seal at once. May have to leave rings on jars to keep tight seal. Takes about 4 to 6 weeks to be ready to use. This is an old recipe and never fails but one thing important is to put cucumbers in same day you pick. This amount of brine covers about 1 gallon, so increase to your needs.

Mrs. Glen Miller
Louisville, Nebraska

Kosher Dill Pickles

Cucumbers
1 cup pickling salt
8 cups water
2 cups white vinegar
3 cups water
2 tbsp. pickling spice
8-12 bunches dill
8 red peppers
8 garlic buttons

Soak 35-40 cucumbers in brine of pickling salt and water. Remove from brine and dry. Combine vinegar, water, spice and dill. Place over high heat and bring to a boil, add cucumbers and remove from heat. Pack cucumbers in jars with at least 1 bunch dill, 2 hot peppers and 2 garlic buttons per jar. Reheat vinegar and pour over pickles. Seal jars.

Mrs. Eliza Stephens
Elm Creek, Nebraska

Kosher Dills

Cucumbers
Dill
1 tsp. pickling spice
3 red peppers
Garlic bud
3 qt. water
1 cup salt
1 tsp. alum

Pack dill-size cucumbers in jars with fresh dill. In each quart jar put pickling spice, peppers and garlic bud. Boil water, salt and alum. Pour hot or cold over pickles and seal. (The small red peppers in pickling spices can be used.)

Mrs. Ven Waddell
Indianola, Nebraska

Kosher Dill Pickles

Cucumbers
1 cup pickling salt
8 cups water
2 cups white vinegar
3 cups water
2 tbsp. pickling spice
Dill
Garlic

Soak 35-40 cucumbers for 24 hours in brine of salt and water - remove from brine and dry. Combine vinegar, water, spice and dill. Bring to boil and add cucumbers. Remove from heat, pack in jars with bunch of dill, add clove of garlic. Place vinegar back on heat and bring to boil. Pour over cucumbers and seal.

Mrs. Orville Bradley
Omaha, Nebraska

Lazy Housewife Sweet Dill Pickles

Cucumbers
Alum
4 cups sugar
¼ cup pickling salt
1 qt. vinegar
1 pt. water
2½ tsp. celery seed

Put dill into sterile jars. Slice cucumbers quite thin lengthwise. Pack in jars. Put a pinch of alum in each jar. Mix sugar, salt, vinegar, water and celery seed and bring to boil. Pour over pickles and seal. They are ready to use in 2 wks.

Mrs. Murl Smith
Defiance, Iowa

Never Fail Dill Pickles

Cucumbers
Dill
½ tsp. pickling spice
Green pepper
Garlic
3 cups water
1 cup vinegar
¼ cup salt

Pack cucumbers as perpendicularly as possible in 1 or 2 qt.jars, with a head of dill, pickling spices and a thin slice of green pepper, also alum. (Garlic is optional) Clover of garlic is optional. Boil water, vinegar, salt and pour over pickles. Seal tightly, If exact directions are followed these pickles will stay crisp a year or longer. Ready to eat in 3 wks.

Mrs. Ed Scheuneman
Ithaca, Nebraska

Mrs. Marvin Mattson
Hooper, Nebraska

Mrs. Otto Schmidt
Arlington, Nebraska

Never Fail Dill Pickles

Cucumbers
1 cup sugar
½ cup pickling salt
4 cups vinegar
4 cups water
Dill heads
Pickling spices
Onions

(Continued on next page)

Wash and slice cucumbers in quarters lengthwise. Put 2 dill heads, ½ tsp. pickling spices and 2 small onions in each quart jar. Pack with cucumbers heat sugar, salt, vinegar and water to boiling and pour over pickles. Seal jars, place in hot water bath kettle until full rolling boil, turn off heat. Leave jars in water 5 min. longer, remove and set on board or cloth to cool. These pickles are ready to eat in a few days.

Ralph J. Thompson
Curtis, Nebraska

Never Fail Dill Pickles

Cucumbers
Dill
Pickling spices
¼ tsp. alum
3 cups water
1 cup vinegar
¼ cup pickling salt

Pack cucumbers in quart or ½ gal. jars. Put a head of dill in the bottom and a few pickling spices on top, if desired. Add ¼ tsp. of alum on top of each quart. Boil water, vinegar and salt; pour over pickles. Add 6-8 heads of dill to ½ gal. jar or 1 tbsp. dill seed to a quart. They are ready to use in 3 wks.

Mrs. Frank Hahn
Johnson, Nebraska

Never Fail Dill Pickles

Cucumbers
Dill
2 tbsp. pickling salt
2 cups water
1 cup vinegar
Garlic cloves

Scrub dill size cucumbers and stand in sterile quart jars leaving enough room for the dill. Pour boiling water over cucumbers to top of jar and let stand until water has cooled to lukewarm. Drain. Boil salt, water and vinegar and pour over pickles as they are in jars. Add 2 cloves garlic and 3 heads of dill for each jar and seal. These are ready in 5 to 6 wks. NOTE: These of course will not ferment or ooze out. We make them each season providing we raise the cucumbers. We have shared this simple recipe with many as they are so good and easy to make.

Mrs. Clara Loberg
Norfolk, Nebraska

Old Fashioned Dill Pickles

Cucumbers
Dill heads
3 qt. water
1 qt. vinegar
1 cup non-iodized salt
Grape leaves

Pack 3-4 stalks of dill heads and stalks in bottom of 2 or 3 gal. crock. Wash dill sized cucumbers and pack in crock over dill. Mix water, vinegar and salt and bring to a boil. Pour over cucumbers in jar and cover with 2-3 grape leaves that have been thoroughly washed. Place an old plate over all and weigh down with a jar filled with warm water. Tie a clean cloth - old dish cloth or such over all and place in a warm or temperate place. After 3 days test the cucumbers by eating a piece of one. It may be ready to eat in 3 days, if not it will be ready in about 5 days. These may be eaten from the crock or if you desire to preserve them, you may pack into jars, bring solution to boil, pour over pickles and seal.

Mrs. Louis Gross
Fremont, Nebraska

One Day Dill Pickles

Cucumbers
Dill
Garlic
13 cups water
6 cups white vinegar
3/4 cup pickling salt
Sugar

Pack jars solid with pickles, dill and a bud of garlic. Boil water, vinegar and salt for 15 min. 1 tsp. sugar may be added to each jar. Pour hot brine in jars and seal snug, not forced down. Take container like used for cold packing, put jars in and fill with boiling hot water to 1 in. from top of lids. Do not have water above lids. Cover container, let stand until next day when water has cooled off. Seal. Pickles are ready for use. Makes enough for 12 qts.

Phyllis Munson
Lincoln, Nebraska

Open Dills In A Crock

Cucumbers
Dill
Grape leaves
10 qt. water
2 cups vinegar
2 cups salt

Wash cucumbers. Alternate layers of cucumbers, dill and grape leaves in crock. Make brine of water, vinegar and salt. Pour on cold - weight down and let set for approximately 1 wk. or longer.

Mrs. Laddie Jirovsky
Hooper, Nebraska

Open Jar Dill Pickles

Cucumbers
3 cups salt
Grape leaves
Dill

Dissolve salt in boiling water. Let cool. In a jar alternate layers of grape leaves, dill and pickles. Cover with brine. Set away 2 wks. Pour brine off if you wish to can, boil and skim. Pack pickles in jars, pour this brine over while still hot, seal. Do not have to can unless you care to.

Mrs. Jay Collins
Alliance, Nebraska

Open Jar Dill Pickles

Cucumbers
Dill
Grape leaves
4 tbsp. sugar
4 tbsp. pickling salt
½ cup vinegar
6 cup water

Put dill and grape leaves on bottom of jar, then put pickles in layers, put dill and grape leaves on top of pickles. Boil sugar, salt, vinegar and water. Let cool and pour over pickles. Weight with plate so that pickles remain immersed in the brine.

Mrs. Mabel Cippera
Morse Bluff, Nebraska

Open Jar Dill Pickles

Cucumbers
Dill
Grape leaves
Horseradish root

9 tbsp. pickling salt
1 cup vinegar
1 gal. water
1 tsp. powdered alum

Wash dill-size cucumbers and pack in layers in 1 or 2 gal. stone jars. Put heads and stems of fresh dill, some grape leaves and chunks of peeled horseradish root between layers. Boil salt, vinegar, water and alum for 5 min. and pour over cucumbers. Garlic cloves may be added to taste. To keep cucumbers under brine, put clean white cloth over cover with plate and weight down with fruit jar filled with water. When at desired taste keep in refrigerator. Make additional brine if needed. It will take from 2-3 wks. to make these.

Mrs. Mary O'Hare
North Bend, Nebraska

Open Jar Dill Pickles

Cucumbers
Dill
Grape leaves
Cherry leaves

1 gal. water
1 pt. vinegar
1 cup pickling salt

Boil water, vinegar and 1 scant cup of pickling salt and pour over cucumbers. Put dill, cherry leaves and grape leaves in bottom of jar. Add layers of pickles with few grape leaves between layers. Finish off with more dill, grape leaves and cherry leaves. Make brine of water, vinegar and salt. Pour hot brine over cucumbers and weigh down with plate. Let stand in fairly warm place for 10 days to 2 wks. Then ready to eat. Makes 2 gal.

Miss Melva Risse
West Point, Nebraska

Open Jar Dills

Cucumbers
Cherry leaves
Grape leaves
Dill

15 cups water
1 cup vinegar
1 cup salt
Alum

Heat water, vinegar and salt to a rolling boil. Cool. Place a layer of dill flowerettes and stems, a few grape leaves and a few cherry leaves - layer of dill-size cucumbers and next the same layer of leaves and dill repeated until jar is full. Each cucumber may be wrapped in a grape leaf. 1 tsp. of alum may be added, if desired. Place these in a 5 gal. crock. Cover cucumbers with cooled brine (if more liquid is needed double the recipe). Place a table plate on top of cucumbers and weight it down with a pint jar of water. Cover with a cloth in warm room.

Frank Peltz, Sr.
Linwood, Nebraska

Open Jar Dills

Cucumbers
Dill

1 gal. water
1 cup pickling salt

Boil water and salt; let cool. Have ready in stone jar, washed cucumbers and dill. Pour the cooled brine over cucumbers, let stand a week or 10 days. Can eat right from the jar or they may be canned. To can strain 6 cups brine, add 1 cup vinegar; boil cucumbers in brine 1 min. Pack in jars, add alum - just a small pinch, a little dill and pour hot brine over pickles and seal.

Mrs. Frank Nickman
Pleasanton, Nebraska

Our Favorite Dill Pickles

Cucumbers
1/8 tsp. alum
1 clove garlic
2 heads dill
Grape leaves

1 qt. vinegar
3 qt. water
1 cup coarse salt

In each quart put alum, garlic, dill and a couple of grape leaves. Pack cucumbers in well but do not force to bruise. Heat to boiling vinegar, water and salt; pour over pickles, seal and set aside for at least 4 wks. before using.

Mrs. Elmer Buchholtz
Morse Bluff, Nebraska

Perfect Dill Pickles

Cucumbers
Dill
1 clove garlic
1 tsp. sugar

13 cups water
1 cup coarse salt
6 cups vinegar

Pack cucumbers in sterilized jars with 1 head of dill and 1 clove of garlic. Pour boiling brine of water, salt and vinegar over cucumbers. Add 1 tsp. sugar on top of each. Seal jars and set in hot water, letting them stand until water is cold. This brine will not ferment. Do not add sugar to brine but sprinkle on top before sealing jars. Eat in 6 wks. or sooner. NOTE: Mature dill heads may be cut, washed and stored in plastic bags in the freezer until needed.

Mrs. Harold Witulski
Beatrice, Nebraska

Mrs. Harvey Wagner,Dorchester, Nebraska sent the same Perfect Dill Pickle recipe as Mrs. Harold Witulski. Mrs. Wagner does not use cloves of garlic in her recipe.

Perfect Dill Pickles

Cucumbers
Dill

13 cups water
1 cup salt
6 cups vinegar

Pack cucumbers in jars with lots of dill. Pour boiling brine of water, salt and vinegar over the cucumbers. Fill to tops of the jars and seal tight. Set jars in boiling water to come over the tops of the jars. Let stand in hot water until it becomes cold. These are crisp so will not ferment and stay clear.

Mrs. Arnold Witt
Lincoln, Nebraska

Pickles

Cucumbers
Dill
Garlic

5 cups water
2 cups vinegar
½ cup salt

Put cucumbers in jar with dill and garlic. Bring water, vinegar and salt to boil and pour over cucumbers. Seal.

Mrs. L. C. Rieger
Lincoln, Nebraska

Dill Pickles

Cucumbers
Fresh dill

3 cups water
1 cup vinegar
¼ cup salt

Bring water, vinegar and salt to a boil. Pour slowly over pickles that are packed in sterilized jars. Add stalk of dill. Use zinc lids and rubber rings. Seal tightly. Set away and let work. Ready to use in 5 wks. This is enough liquid for 2 qts.

Mrs. Wm. J. Murphy
O'Neill, Nebraska

Plain Dill Pickles

Cucumbers
Dill
3 qt. water

1 qt. vinegar
1 cup pickling salt
3/4 tsp. alum

Pack washed pickles in a jar. Add some dill heads, 2 or 3 to a quart jar, more to a larger jar. Boil water, vinegar, salt and alum and pour over pickles in jars and seal. For a sweet pickle, add 2 cups sugar and also can put a slice of onion in the bottom of jar with the dill and proceed as before boiling the brine and pouring hot over pickles. One can slice the pickles or cut lengthwise, whatever is preferred. Garlic may be added.

Mrs. Frank Nickman
Pleasanton, Nebraska

Plain Hamburger Dills

Cucumbers	1 cup salt
Dill	5 cups vinegar
Grape leaves	7 cups water
	1 tsp. alum

Boil salt, vinegar, water and alum and pour over dill-size cucumbers. Add dill and grape leaf to each jar and seal.

Mrs. E. C. Volkmer
Talmage, Nebraska

Quick Dill Pickles

Cucumbers	1 part water
Pickling salt	1 part sugar
Dill	1 part vinegar

Wash and cut cucumbers (all sizes) into chunks. Sprinkle a little pickling salt over them and let stand till you prepare the brine. Make brine of water, sugar and vinegar; bring to boil. Cook cucumbers in brine until they lose their bright green and turn a dull green. Pack in jars. Place a head of dill on top of each jar and seal.

Mrs. Edward Havlat
Crete, Nebraska

Mrs. Harvey Wagner
Dorchester, Nebraska

Quick Dill Pickles

Cucumbers	2 qt. water
Dill	1 qt. vinegar
Garlic cloves	1 cup salt

Wash and use cucumbers as soon as possible after picking. Use 2 quart or gallon jars. Put a couple sprigs of dill in the bottom of the jar and pack cucumbers in, add another sprig or two of dill, then more cucumbers and some more dill on top. Garlic may be added to each jar, if desired. Boil water, vinegar and salt. While still boiling pour over cucumbers and seal at once. This makes an excellent garlic-dill flavor which produces a pickle very much desired for hamburgers and snacks.

Mrs. John W. MacDonald

Ruby's Sweet Dills

Cucumbers	1 cup vinegar
Dill	1 cup sugar
1 cup salt	3/4 cup water
1 gal. water	

Wash and split cucumbers - or use whole if not too large. Soak cucumbers in brine of salt and water overnight. Next morning drain and drop in boiling syrup of vinegar, sugar and water. Then heat in kettle until change color. Put dill in bottom of jars, add pickles and brine; seal while hot.

Mrs. Frank J. Hobbs
Lincoln, Nebraska

Russian Dills

Cucumbers
Onion
Dill

2 cups vinegar
1 cup water
1½ cups sugar
¼ cup pickling salt

Wash and slice thin circle of cucumbers into quart jars, slice one onion into each jar. Put a head of fresh dill on top and pour boiling solution of vinegar, water, sugar and salt over them. Seal. No work with this recipe, and they are delicious.

Mrs. Joe Novak
Dwight, Nebraska

Short Method - Dill Pickles

Cucumbers
Green or dry dill
3/4 cup sugar
½ cup salt

1 qt. vinegar
1 qt. water
3 tbsp. mixed pickling spice

Cut 30-40 medium cucumbers in half lengthwise. Combine sugar, salt, vinegar and water. Tie spices in a cloth bag. Add to vinegar mixture: Simmer 15 min. Pack cucumbers into hot jars, leaving ¼ in. head space. Put head of dill in each jar. Heat brine to boiling. Pour boiling hot vinegar mixture over cucumbers leaving ¼ in. head space. Adjust caps. Process pints and quarts 15 min. in boiling water bath. Yields about 7 pt.

Mrs. Frank J. Hobbs
Lincoln, Nebraska

Sour Dills

Cucumbers
Dill
1 cup salt

3 cups vinegar
9 cups water

Wash cucumbers well and pack in sterile jars with dill as desired. Make a brine of salt, vinegar and water. Pour boiling over pickles in jars and seal. Ready in a few weeks.

Mrs. Ralph Culp
Crete, Nebraska

Spanish Dill Pickles

Cucumbers
Dill
1/3 can pimientos or green peppers

2 qt. water
1½ qt. vinegar
Scant cup salt
2 cups sugar

Sterilize jars and fill with lengthwise sliced cucumbers. Add dill and 1/3 can pimientos or green peppers. Set filled jars in hot water and slowly pour the following liquid over the cucumbers and seal tightly: water, vinegar, salt and sugar. Makes about 8 qt.

Spirit Lake, Iowa

Sweet Dill Pickle

Cucumbers
Dill
1 qt. vinegar
1 cup water
1 cup sugar
½ cup salt (canning salt)

Soak cucumbers 1 hr. in very cold water or ice water. Pack in jars with piece of dill in bottom and top of jar. Pour boiling mixture of vinegar, water, sugar and salt over pickles and seal. Makes about 3 qt. P.S.: I put a little alum in each jar to keep them crisp.

Mrs. J. B. Teeters
Decatur, Nebraska

Sweet Dill Pickle

Cucumbers
Dill
Garlic
Onion
4 cups water
2 cups vinegar
1 cup sugar
½ scant cup salt

Wash pickles. Put small cut onion, 1 tsp. dill and garlic cut in bottom of jar. Pack in quartered pickles. Then add couple small pieces of garlic, onion and ¼ tsp. dill on top. Jar to be filled with boiled mixture, then sealed. Place jars in hot water and let stand all day. If sunny day put on sidewalk.

Mrs. M. Buschkamp
Lincoln, Nebraska

Sweet Dill Pickle

Cucumbers (dill size)
3-4 slices onion
Dill
5 cups sugar
5 cups light vinegar
1/3 cup pickling salt
1 tsp. alum

Slice dill-size cucumbers with skin on lengthwise. Slice quite thin. Pack in jars. Place 3 or 4 slices onion and dill on top. Heat to boiling the syrup of sugar, vinegar, salt and alum. Pour hot syrup over pickles. Seal and cold pack just until water in canner boils. Remove from canner and let stand inverted until cold.

Mrs. Henry Paeper
Nickerson, Nebraska

Sweet Dill Pickle

Cucumbers (dill size)
White onion
2 celery stalks (quartered)
8 heads fresh dill
1 qt. vinegar
4 cups sugar
2 cups water
½ cup salt

(Continued on next page)

Wash freshly picked cucumbers and cut into 1-in. chunks or quarters or slice. Fill 4 sterilized jars with cucumbers. To each jar add 3 or 4 slices onion, 2 pieces celery and 2 heads of dill. Dissolve sugar and salt in vinegar and water, bring to a boil. Pour hot solution over cucumbers in jars and seal. Store in dark cool place for 30 days. They work in the jars so do not think they are spoiling.

Miss Adelhied Hilmer
St. Paul, Nebraska

Sweet Dill Pickle

Cucumbers	3 cups water
3 cups sugar	1/3 cup salt
3 cups vinegar	Dill

Boil the water, vinegar and sugar, pour over pickles in a stone jar.

Bessie Meduna
Weston, Nebraska

Sweet Dill Pickle

1 gal. cucumbers	1 tbsp. alum
3/4 cup coarse salt	1 tbsp. turmeric
2 pt. vinegar	Dill
4 pt. water	4 cups sugar

Cut cucumbers into chunks. Use coarse salt and boiling water. Let stand overnight. In morining drain. Simmer ½ hr. in 1 pt. vinegar, 3 pt. water, 1 tbsp. alum 1 tbsp. turmeric. Drain. Pack pickles in jar with dill. Then heat 1 pt. vinegar, 1 pt. water, 4 cups sugar. Boil and cover pickles. Seal while hot. Let stand several weeks before using.

Mrs. Ed Scheuneman
Ithaca, Nebraska

Sweet Dill Pickles

Cucumbers	3½ cups white vinegar
Dill	3 cups water
Hot pepper	¼ cup pickling salt
Garlic	½ cup sugar
Onion	½ tsp. alum
	½ tsp. turmeric

Wash and cut cucumbers in 4ths lengthwise. Put dill in bottom of jar, add strips, put in a piece of hot pepper, garlic and onion and finish the top of jar with dill. Bring brine of vinegar, water, salt, sugar, alum and turmeric to good boil, pour over pickles and seal with hot lids.

Mrs. Marie Supanchick
North Platte, Nebraska

Sweet Dill Pickles

Bought dill
9 pickles
½ tsp. celery seed
2 cups sugar
1 tsp. mustard

Put into jars and eat when you feel like it.

Mrs. Ivar G. Saf
Oakland, Nebraska

Sweet Dill Pickles

Cucumbers
1 qt. cider vinegar
1 pt. water
½ cup salt (pickling)
1 tsp. alum
Fresh dill

Bring vinegar, water, sugar, salt and alum to a boil. Use only fresh cucumbers about length of a quart jar. Cut cucumbers in 1/8ths lengthwise after washing. Pack tightly in quart jars with a large piece of dill to each quart. Cover with boiling liquid and seal. Let stand 3-4 wks. before using. The above quantity make about 4 qt. pickles. If there is liquid left, I seal it in a jar and refrigerate to use again, reheating it of course. This does not make a candied-type dill, just a sweet-sour dill. It is particularly good to use when you have only a few cucumbers at a time, as it is so simple.

Mrs. D. W. Moench
Omaha, Nebraska

Sweet Dill Pickles

1 gal. dill-size cucumbers
3/4 cup salt
2 pt. vinegar
4 cups sugar
4 pt. water
1 tbsp. alum
1 tbsp. turmeric

Slice cucumbers. Cover with 3/4 cup salt and enough boiling water. Let stand overnight. Drain in morining, rinse, then simmer ½ hr. in 1 pt. vinegar, 3 pt. water, 1 tbsp. alum, 1 tbsp. turmeric. Then take pickles out of solution and pack in sterilized jars, and fill each jar with syrup of 1 pt. vinegar, 1 pt. water, 4 cups sugar. Boil together, then pour over pickles and seal.

Mrs. Wm. J. Murphy
O'Neill, Nebraska

Sweet Dill Pickles

Cucumbers (dill sized)
3 cups vinegar
3 cups sugar
2 tbsp. mustard seed
2 tbsp. dill seed
2 tbsp. salt
¼ tsp. alum

(continued on next page)

Slice dill sized pickles ¼ in. thick. Slice a full rounded gallon crock full. Cover with cold water and let stand a few hours until crisp. Simmer in brine of vinegar, sugar, mustard seed, dill seed, salt and alum 15 min. Do not boil, seal while hot. Very good for hamburgers or hot dogs.

Mrs. J. W. Glock
Ulysses, Nebraska

Sweet Dills

Cucumbers
Onion
Dill
1 qt. vinegar
1 pt. water
½ cup salt
2/3 cup sugar

Cut your cucumbers in halves or quarters. Place sliced onion on top. Then cover with hot vinegar as follows: 1 qt. vinegar, 1 pt. water, ½ cup salt, 2/3 cup sugar. Heat and pour over cucumbers and seal. Ready to eat in 3 wks.

Mrs. William McKinzie
Beatrice, Nebraska

Sweet Dills

1 gal. cucumbers
3/4 cup salt
2 pt. vinegar
3 pt. water
1 tbsp. alum
1 tbsp. turmeric
Dill
4 cups sugar

Cover cucumber chunks with 3/4 cup salt and boiling water and let stand several hours. Remove from brine and simmer in 1 pt. vinegar, 3 pt water, alum and turmeric until rinds turn color. Remove from this and pack into hot, sterile jars. Add dill to taste and cover with hot syrup of 1 pt. vinegar, 4 cups sugar. Seal.

Mrs. Lawrence Quante
Brock, Nebraska

Sweet Dills

Cucumbers
1 cup pickling salt
2 cups water
1 cup sugar
1 cup vinegar
pinch of alum
Dill

Put whole medium or dill-size cucumbers in a brine of 1 cup pickling salt to 1 gal. cold water, dissolved. Soak no longer than 2 days. Heat 2 cups water, 1 cup sugar 1 cup vinegar to boiling point, add a good sized pinch of alum. Split cucumbers in half lengthwise and put in boiling syrup only to boiling point. Take out immediately and put in hot jars with a head of dill in bottom and one on top and fill with the hot syrup and seal. When using, cut in finger lengths or slices as desired.

Mrs. Ralph Culp
Crete, Nebraska

Sweet Dills

1 gal. cucumbers
3/4 cup salt
1 pt. vinegar
3 pt. water
1 tbsp. alum
1 tsp. turmeric
Dill

Cut 1 gal. cucumbers in chunks (or split smaller ones). Add 3/4 cup salt and cover with boiling water. Let stand overnight. Then drain and rinse in clear water--simmer ½ hr. in 1 part vinegar to 3 parts water, 1 tbsp. alum and 1 tsp. turmeric. Pour off and discard--pack jars with sprig of dill and cover with boiling syrup made of 1 cup vinegar and 1 cup sugar.

Mrs. Orville Bradley
Omaha, Nebraska

Sweet Sour Dills

Cucumbers (dill)
Onion
Dill
¼ tsp. of alum
1 cup sugar
1 cup vinegar
1 cup water
2 tbsp. pickling salt

Cut dill cucumbers into 4th lengthwise, put one head of dill and a slice of onion in bottom of jars, pack cucumbers in firmly, put another slice of raw onion on top and a small head dill, ¼ tsp. alum on top also. Boil and pour hot the solution of sugar, vinegar, water, and pickling salt over the cucumbers.

Mrs. Charles Weber
Columbus, Nebraska

Sweet-Sour Dill Chunks

Cucumbers (dill size)
Onion slices
2 celery stalks (quartered) (optional)
8 heads fresh dill
4 cups sugar
½ cup salt
1 qt. vinegar
2 cups water

Wash freshly picked cucumbers and cut in 1 in. chunks or in quarters enough to fill 4 sterlized quart jars. To each jar add 3 or 4 slices onion, 2 pieces celery and 2 heads dill. Dissolve sugar and salt in vinegar and water; bring to a boil. Pour while hot over cucumbers in each jar to cover; seal at once. Store in a cool place. For best flavor, do not use for 30 days. Variations: slice cucumbers ¼ in. thick. Just right to tuck in hamburgers.

Mrs. A. Sell
Milford, Nebraska

1-3-9 Dill Pickles

Cucumbers
1 cup pickling salt
9 cups water
3 cups cider vinegar
Dill
Garlic
Onions (pickling)

(Continued on next page)

Stir and bring to a boil the salt, water and vinegar, so salt will dissolve. Pack clean, fresh, pickling size cucumbers into sterilized qt. jars. Put 2 heads of dill into each jar. Bring salt, vinegar and water to boil and while boiling hot fill the jars of cukes to top and seal with zinc lids and rubbers. Seal tight. Let stand in a warm place about 10 days until fermentation ceases, then store in a cool basement. You may add a clove of garlic or pickling onions for variation.

Mrs. Harvey Wagner
Dorchester, Nebraska

Sweet Sour Dill Chunks

Cucumbers (dill size)
Onion slices
2 celery stalks (quartered)
8 heads fresh dill

4 cups sugar
½ cup salt
1 qt. vinegar
2 cups water

Wash freshly picked cucumbers and cut in 1 in. chunks enough to fill 4 qt. jars. To each jar add 3 or 4 slices of onion, 2 pieces of celery and 2 heads of dill. Dissolve sugar and salt in vinegar and water, bring to a boil and pour over cucumbers in each jar to cover. Seal. Store in cool place and do not use for 30 days. Good.

Ivy Lawton
Raymond, Nebraska

Unusual Dills

12 lb. cucumbers
Dill
Grape leaves
Garlic
1 tsp. pickling spice

6 qt. water
1 tsp. powdered alum
3 cups salt
1 qt. vinegar

Soak cucumbers (3-4 in. long) overnight. Next morning scrub them carefully and pack in 2 qt. jars. In the bottom of jar put a liberal amount of dill. I use green dill, leaves of grapes and a small clove of garlic. Place 1 tsp. of pickling spices, the king with small red pepper in it. Then fill the jars ½ full of cucumbers, put in a small clove of garlic and a very liberal amount of dill and leaves and another small clove of garlic and 1 tsp. of pickling spices. Be sure to use plenty of dill. Put to boil the water, add a tsp. of powdered alum and 3 cups salt. When mixture boils, remove scum that rises to the top. When boiling point is reached, add a qt. of vinegar and bring to boiling point again. Fill jars with liquid, expelling all air and seal immediately. It is necessary to seal these pickles in airtight jars. Makes approximately 12 qt. pickles.

1-3-9 Dill Pickles

Cucumbers
Dill
1 cup salt

3 cups vinegar
9 cups water

This famous recipe has stood the test of years. Arrange cucumbers in 2 qt. jars. Place dill in the bottom, center and top of jar. Boil mixture of salt, vinegar and water. Fill jars and seal.

Mrs. Lewis Royal
Nehawka, Nebraska

1-2-9 Dill Pickles

Cucumbers
Dill
1 cup pickling salt
2 cup brown vinegar
9 cups water
¼ tsp. alum

Fill jars with pickles, dill size. Put dill on top. Boil together the salt, brown vinegar, water and alum. Pour boiling hot over pickles and seal immediately. Set in a dark place. Makes 4 qts.

Mrs. Ed Scheuneman
Ithaca, Nebraska

HEINZ PICKLES

Heinz Pickles

Cucumbers
Pickling salt
2 qt. vinegar
1 tsp. celery seed
1 tsp. mixed spices
A little stick of cinnamon

Put cucumbers in brine made of a cup of pickling salt to each gallon of water used. Let stand 24 hr., drain, cover with clear, boiling water. Let stand 24 hr., drain and split each pickle. Then cover with boiling hot liquid made from vinegar, 8 cups sugar, celery seed, mixed spices and cinnamon. Let stand 24 hr., pour off and reheat 3 mornings without adding sugar. Last time put hot pickles in jars and pour over the hot liquid and seal.

Mrs. Charles Petersen
Cook, Nebraska

Heinz Pickles

2 gal. cucumbers
1 cup salt
1 gal. water
3 tbsp. alum
8 cups vinegar
10 cups sugar
2 tsp. celery seed
2 tsp. mustard seed
¼ box of whole cloves
1 box stick cinnamon

Soak cucumbers 7 days in salt and water mixture. Drain, then soak 2 days in 1 gal. boiling water containing alum. Drain and wipe each pickle dry with cloth, then split pickles. Heat vinegar, sugar, celery seed, mustard seed, cloves and cinnamon and pour over pickles 3 days in succession, draining and reheating liquid. This is delicious pickle, especially if made with cucumber.

Mrs. Audra Halbert
Falls City, Nebraska

Heinz Sliced Sweet Pickles

Medium and large cucumbers
Coarse pickling salt
Sugar
Alum
2 cups vinegar
1 oz. whole allspice
1 oz. stick cinnamon (or 1 oz. cassia buds)
3 tsp. celery seed

Slice cucumbers thin. Cover with brine (strong enough to float an egg) made by using pickling salt and water. Let soak 3 days, then drain off salt water and put pickles in fresh, cold water 3 days (changing the water each day). Drain and simmer, not boil, pickles for 2 hr. in mixture of 6 cups water, 1 cup sugar and alum size of walnut. Then drain off this liquid and throw away. Then make liquid of 5 cups sugar, vinegar,allspice, cinnamon (or cassia buds) and celery seed, and bring pickles to boil, can and seal. Allspice cinnamon (or cassia buds) and celery seed should be tied in a thin cloth bag. These are real good pickles,sweet and crisp.

Mrs. Homer C. Blakeman
Merna, Nebraska

Heinz' Pickles (Sweet)

100 pickles
1 pt. salt
1 tbsp. powdered alum
9 cups sugar
5 pt. white vinegar
1 oz. celery seed

Start with about 100 pickles 3 to 5 in. long, wash and drain. Cover with a brine of 1 gal. cold water and 1 pt. salt, let stand 1 wk., skim if necessary. Then drain and cover with boiling water to which add 1 tbsp. powdered alum, let stand 24 hr. Drain and split in half (regardless how small), cover with boiling water again and let stand 24 hr. again. Then drain and cover with syrup made of 6 cups sugar, vinegar and celery seed, boiling hot. Let stand 1 wk. and each morning for the first 3 days drain, heat, add 1 cup sugar and pour back on pickles. After 3 days heat syrup without adding any more sugar and pour back on pickles. End of week pack in sterilized jars, heat same syrup, pour over them and seal.

Mrs. J. F. Janousek
Lincoln, Nebraska

Heinz Sweet Pickle

Cucumbers
Salt
Alum
2½ gal. vinegar
Sugar
1 stick cinnamon
1 tbsp. celery seed
1 tsp. whole cloves

Wash 1 bu. small cucumbers and pack in crock. Cover with brine made of 1 pt. salt to 1 gal. water. Let stand 1 wk. Pour off and cover with solution made of 1 tbsp. alum to 1 gal. water, heated to boiling. Let stand 24 hr. Drain and cover pickles with clear boiling water and let stand 24 hr. Pour off liquid. Heat vinegar, cinnamon, celery seed, cloves and 1 cup sugar and pour over pickles. Let this stand 3 days. Pour vinegar solution in kettle, add 1 cup sugar, heat to boiling and pour over pickles. Repeat until 3 cups sugar have been added to original solution. Let stand 3 days between each addition. Reheat vinegar if you desire to seal. These pickles are clear and keep their green color.

Mrs. Leon J. Thiahrt
Rapid City, So. Dakota

Heinz Sweet Pickles

Cucumbers
Coarse pickling salt
Alum
2½ lb. sugar
1 qt. vinegar
1 tbsp. mixed spices

Soak cucumbers in large stone jar in salt water strong enough to float an egg. Use coarse pickling salt. Soak 7 days. Then pour liquid off and soak in fresh water 2 days. On morning of second day cut cucumbers lengthwise. Bring to boil in alum water (2 tbsp. alum to 1 gal. water). Let them stand ½ day. Then boil sugar, vinegar and spices (in bag). Pour syrup over pickles. (Green color may be added to syrup if desired). Drain off syrup and heat again. Pour over pickles 3 mornings (as above). Seal on third morning. May want to add more green coloring.

Mrs. Lela Daily
Anselmo, Nebraska

Heinz Sweet Pickles

Cucumbers
Salt
Grape leaves
Vinegar
Granulated sugar
Mixed whole pickling spices
Alum

Slice lengthwise 2 gal. cucumbers. Soak 3 days in salt solution (2 cups salt to 1 gal. water) then freshen 2 days, changing water each morning. Lay layer of grape leaves, then layer of cucumbers alternately in large kettle and boil in weak vinegar solution (1 pt. vinegar to 1 gal. water). Cook very slowly 1 hr. Grape leaves restore fine green color to pickles. Pack cucumbers in jars and pour over pickles a syrup made of 1 qt. vinegar, 3 lb. granulated sugar, mixed whole pickling spices and 2 scant tbsp. alum. Seal. These are very delicious.

Box 244
Hooper, Nebraska

Heinz Sweet Pickles

75 medium-size cucumbers
Coarse pickling salt
Alum
5 pt. vinegar
White sugar
1 tsp. celery seed
1 tbsp. whole allspice
Some stick cinnamon

Cover cucumbers with brine made of 1 cup coarse pickling salt to each gallon water in stone jar or crock. Soak cucumbers in brine 7 days. Take out of brine and split cucumbers lengthwise. Cover cucumbers with boiling alum water (1 tbsp. alum to each gallon water). Let stand 24 hr. and drain. Make syrup of vinegar, celery seed allspice, cinnamon and 6 cups sugar, boil until sugar dissolves. Pour boiling syrup over drained cucumbers. Next morning drain syrup, add 1 cup sugar and boil until sugar dissolves. Pour over cucumbers. Repeat process 5 days, adding 1 cup sugar to drained syrup, heating and pouring over pickles. Pickles now ready to eat or can.

Mrs. Clair Blezek
Randolph, Iowa

Heinz Sweet Pickles

1 bu. small cucumbers
Salt
Alum
2½ gal. vinegar
Sugar
1 stick cinnamon
1 tbsp. celery seed
1 tsp. whole cloves

Wash and pack cukes in crock. Cover with brine made of 1 pt. salt to 1 gal. water. Let stand 1 wk. Pour off and cover with solution 1 tbsp. alum to 1 gal. water, heated to boiling. Let stand 24 hr. Drain and cover pickles with clear boiling water and let stand 24 hr. Pour off liquid. Heat vinegar, cinnamon, celery seed, cloves and 1 cup sugar and pour over pickles. Let stand 3 days. Pour vinegar solution in kettle, add 1 cup sugar, heat to boiling and pour over pickles. Repeat until 3 cups sugar have been added to original solution. Let stand 3 days between each addition. Reheat vinegar if you desire to seal.

Mrs. Leon Thiahrt
Rapid City, So. Dakota

LIME PICKLES

Blue Ribbon Lime Pickles

2 cups lime
8 cups vinegar
9 cups sugar
1 tsp. mixed pickling spices
2 gal. water
1 gal. cucumbers, sliced
1 tbsp. salt
Green food coloring

Dissolve lime in water. Add sliced cucumbers. Let stand 24 hr. with a plate weighted on top to keep any from floating. (I use a pint jar filled with water or a plastic bag filled with water as the "weight"). After 24 hr., pour off and discard limewater. Wash cucumber slices through 3 waters. Return to jar and cover with cold water. Let stand 3 hr., no longer. Drain. Make a syrup of vinegar, sugar, salt and spices tied in a bag. Bring to boil. Pour over drained pickles. Let stand overnight. Put pickles and syrup into kettle. Boil for 35 min. Add green food coloring until a nice green color is achieved. Ladle hot into sterilized jars and seal. This is an old recipe which has been used by many through the years. It won first place at the Fremont Co. Fair in Sidney, Iowa, last summer so it ranks very high. Since giving this on radio, several suggestions for variations have come in: add 1 tsp. celery seed and 1 tsp. whole allspice to the 1 tsp. pickling spice in the bag for a more pronounced and equally delicious pickle.

Esther Bruneau
Jefferson, So. Dakota

Cucumber Pickles (Ready to eat in 3 days)

8 lb. medium or small cucumbers
2 cups Snowflake hydrated lime
2 gal. water
2 qt. vinegar (or less, according to taste)
8 cups white sugar
1 tsp. salt
1 tsp. celery seed
1 tsp. whole cloves
1 tsp. mixed pickling spices

Slice cucumbers in ¼ in. slices or split small ones lengthwise. Pour a solution of 2 cups lime and 2 gal. water over them and soak 24 hr. Pour off limewater, rinse well, cover with fresh cold water. Let soak 3 hr. Drain, add sugar, vinegar, spices and enough water to cover cucumbers. Let stand overnight. In morning boil gently for 30 to 40 min. Add several drops of green food coloring. Pack in sterilized jars and seal.

Mrs. Joe Zimmerman
Neola, Iowa

Sent in by Mrs. Frank B. Deavers
David City, Nebraska

Evelyn's Cucumber Sticklers

7 lb. cucumbers
2 cups hydrated lime
2 gal. water
2 cups cider vinegar
4½ lb. sugar (equal to 9 or 10 cups)
1 tsp. pickling spice
1 tsp. celery seed
1 tbsp. salt
Green coloring

Use large, peeled cucumbers, seeded and cut into sticks. Put cucumbers in mixture of lime and water. Soak 24 hr. Drain well, Rinse well, cover with clear water. Let set 3 hr. or longer, then drain. Add cider vinegar, sugar, pickling spice, celery seed, salt and coloring. Let stand in pickle brine and let stand overnight. Bring to boil and let simmer 35 min., then can.

Mrs. R. F. Evans
Hickman, Nebraska

Lime Cucumber Pickles

7 to 10 lb cucumbers
2 cups hydrated lime
2 gal. cold water
8 cups sugar
2 qt. vinegar
1 tsp. celery seed
1 tsp. cloves
1 tsp. mixed pickle spices

Wash cucumbers. Slice like bread and butter pickles. Mix water, lime, add cucumber slices, soak 24 hr., stirring once in awhile. Rinse 3 times in cold water. Let soak 4 hr. in fresh cold water. Drain. Make syrup by mixing sugar, vinegar, salt, spices. Pour over cucumbers. Let stand overnight. Next day simmer 35 min., put in jars. Seal. These are good crisp pickles.

Mrs. Edwin Bader
Burr, Nebraska

Lime Cucumber Pickles

7 to 10 lb cucumbers
2 cups air-slaked lime
2 gal. cold water
8 cups sugar
1 tbsp. salt
2 qt. vinegar
1 tsp. celery seed
1 tsp. cloves
1 tsp. pickling spice

Slice cukes. Mix water, and lime. Add slices, soak 24 hours, stirring occasionally. Rinse 3 times in cold water. Soak 4 hr. in fresh cold water, drain. Make syrup of sugar, salt, vinegar, spices. Pour over cukes. Let stand overnight. Next day, simmer 30 min. Seal in hot jars.

Dorothy Straw
Lincoln, Nebraska

Lime Cucumber Pickles

7 to 10 lb. cucumbers
2 cups air-slaked lime
2 gal. cold water
8 cups sugar
1 tbsp. salt
2 qt. vinegar
1 tsp. celery seed
1 tsp. cloves

(Continued on next page.)

Wash cucumbers and slice crosswise as for bread and butter pickles. Mix water, lime, add cucumber slices. Soak 24 hr. Stir occasionally. Rinse 3 times in cold water and soak 4 hr. in fresh cold water. Drain. Make syrup by combining sugar, vinegar, salt, spices. Pour over cucumber slices. Let stand overnight. Next day simmer 35 min. and can in clean hot jars. Seal.

Mrs. Henry H. Frerichs
Minden, Nebraska

Limed Pickles

7 lb. medium-size cucumbers, sliced
2 gal. water
2 cups air-slaked lime
2 qt. white vinegar
8 cups sugar
2 tsp. salt
1 tsp. mixed pickling spice
1 tsp. cloves
1 tsp. celery seed

Soak cucumbers in lime and water 24 hr. Then wash well in several waters (cold). Then soak 3 hr. in clear water. Make syrup of vinegar, sugar, salt, pickling spice, cloves and celery seed. Heat all together and add pickles. Let stand overnight. In morning, let simmer 30 min. and can. This makes very crisp pickles.

Mrs. Carl Richert
Murdock, Nebraska

Lime Pickles

7 lb. medium-sized cucumbers
(thick-slice or cube)
2 cups lime
2 gal. water
2 qt. vinegar
4½ lb. white sugar
1 tbsp. celery seed
1 tbsp. whole cloves
1 tbsp. mixed spice
1 tbsp. salt
1 cup raisins

Soak cukes in lime and water 24 hr. Rinse well. Cover with clear cold water, let stand 3 hr. Drain and cover with vinegar, sugar, celery seed, whole cloves, spice and salt. Let soak overnight. Next morning boil 35 min., then add raisins and boil 5 min. more. Pack hot in sterile jars. (8-10 pt.). Doesn't need processing.

Fern Young
Brock, Nebraska

Lime Pickles

7 lb. cucumbers
Lime
9 cups sugar
2 qt. vinegar
1 tsp. mixed pickle spices
1 tsp. salt
1 tsp. celery seed
1 tsp. allspice (whole)

(Continued on next page)

Slice cucumbers ¼ in. thick. Soak in lime water 24 hr. (2 cups lime to 3 gals. water). Rinse well. Soak 3 hr. in clear water. Drain, cover with 9 cups sugar, vinegar, spices, salt, celery seed and allspice. Let stand overnight, then bring to boil. Boil 30 min and seal in hot jars.

Mrs. George Horner
Lincoln, Nebraska

Lime Pickles

7 lb. sliced cucumbers
2 cups lime
4½ cups sugar
2 qt. vinegar
1 tbsp. salt
1 tsp. celery seed
1½ whole cloves
1 whole cinnamon stick

Let cukes stand 24 hr. in 2 gal. water containing the 2 cups lime. Wash well. Soak in clear water 3 hr. Drain. Mix sugar, vinegar, salt, celery seed, cloves and cinnamon stick and bring to boil. Pour sugar and spices over cucumbers. Let stand overnight. Bring to boil 35 min. Seal. (I soak these in an earthen jar in the lime solution).

Mrs. George J. Pool
Lincoln, Nebraska

Lime Pickles

10 lb. cucumbers
2 cups slaked lime
1 qt. vinegar
1 qt. water
1 tbsp. celery seed
1 tbsp. mixed pickling spice
9 cups sugar
1 tbsp. allspice
1 tbsp. salt

Slice cucumbers. Cover with solution made of slaked lime and 2 gal. water. Soak 24 hr. Then wash thoroughly and cover with cold water and soak 3 hr. longer. Drain and cover with mixture of vinegar, water, celery seed, pickling spice, sugar, allspice and salt. Let stand overnight. In morning, drain and heat liquid to a boil. Put in cucumbers and boil slowly 35 min. Pack in jars and seal.

Lester F. Whitney
Holdrege, Nebraska

Lime Pickles

7 lb. cucumbers
2 cups slaked lime
9 cups sugar
2 qt. white vinegar
1 tsp. salt
1 tsp. mixed spice
1 tsp. celery seed
1 tsp. whole allspice

Wash cucumbers and cut 4 slices to inch. Saok in 3 gal. water with 2 cups slaked lime. Soak overnight or 24 hr. Remove and soak in clear water 3 hr., then drain. Make syrup of sugar, vinegar, salt, spice, celery seed and allspice. Bring to boil. Add cucumbers. Simmer 35 min. If desired you may let them sit overnight and reheat and can.

Mrs. Henry Wendt
Elmwood, Nebraska

Lime Pickles

7 lb. cucumbers
Salt
Household lime
4½ lb. sugar
1 tbsp. mixed spices
1 tbsp. celery seed
2 qt. vinegar

Soak cucumbers in brine 2 wk. Soak in plain water 2 days. Mix 2 cups household lime in 2 gal. water. Slice cukes (thin). Let stand in limewater 24 hr. Drain and wash off lime. Stand in cold water 3 hr. Mix sugar, spices, celery seed and vinegar. Boil. Pour over cuke slices. Next morning, simmer 35 min. in this mixture. Do not let them boil. Seal while hot. Very crisp and green (I also slice cukes in half lengthwise and use same method).

Mrs. W. L. Kinghorn
Lewiston, Nebraska

Lime Pickles

7 lb. sliced pickles
(about ¼ in. thick)
1½ cups slaked lime
2 gal. cold water
2 qt. vinegar
9 cups sugar
1 tsp. celery seed
1 tbsp. whole cloves
1 tsp. whole allspice
1 tbsp. salt
1 tbsp. mixed pickling spice
Green color (if desired)

Soak pickles in limewater 12 hr. Rinse well then cover with clear, cold water. Let stand 3 hr. Rinse again at least once, then cover with mixture of vinegar, sugar, celery seed, cloves, allspice, salt and spice. Let stand in this mixture overnight. Bring to boil, then simmer 35 min. Seal in jars. (If desired, add green color in the vinegar).

Mrs. E. H. Williams
Norfolk, Nebraska

Lime Pickles

7 lb. sliced cucumbers
2 cups ordinary lime
2 gal. cold water
2 qt. vinegar
4½ lb. sugar
1 tsp. celery seed
1 tsp. whole pickling spices
1/3 cup salt

Soak cucumbers in lime solution 24 hr. Rinse 3 or 4 times, then soak 3 hr. longer in clear cold water. Drain. Cover with syrup solution (overnight) made of vinegar, sugar, celery seed, pickling spices and salt. Put pickles and solution in kettle and simmer 30 min. Can and seal while hot.

Mrs. E. B. Staman
Scottsbluff, Nebraska

Lime Pickles

1 gal. fresh cucumbers
2 cups lime
2 gal. water
8 cups vinegar
9 cups sugar
1 tbsp. salt
1 tbsp. celery seed
1 tsp. mixed spices
1 tbsp. whole allspice

Dissolve lime in water. Wash and slice cukes. Put in large crock. Pour lime solution over cukes. Weigh down with plate so solution covers all cukes. Let stand 24 hr. Drain cukes. Wash in fresh water 3 times. Then let cukes stand in cold water 2 hr. (No longer than 3 hr.). Drain cukes. Make syrup for pickles of vinegar, sugar, salt, celery seed, spices and allspice. Bring syrup to boil, pour over cukes. Let stand overnight. Next day put all in big kettle. Boil slowly 35 min. Pack hot in hot jars. Seal immediately. (Do not boil vigorously; pickles will be tough!).

Mrs. Mildred Beyer
Wahoo, Nebraska

Lime Pickles

Medium-size cucumbers
2 cups slaked lime
2 gal. water
2 qt. vinegar
4½ lb. sugar
1 tsp. whole cloves
1 tsp. celery seed
1 tbsp. pickling spices
1 tbsp. salt

Slice cucumbers into large crock or an enamel or granite container. Cover with solution of lime and water. Let stand 24 hr. Drain, wash well. Make syrup of vinegar, sugar, cloves, celery seed, pickling spices and salt. Bring this to boiling point and pour over cucumbers. Let stand overnight. In morning, bring whole thing to boil and simmer 35 min. Pack in hot sterilized jars and seal.

Earl May Seed & Nursery
Shenandoah, Iowa

Lime Sweet Pickles

7 lb. cucumbers, slices or chunks
2 gal. water
1½ cups lime
Sugar
2 tbsp. salt
1 tsp. celery seed
1 tsp. ground cloves
1 tbsp. mixed pickling spices
1 gal. white vinegar
Green food color (if desired)

Get those "cukes" in the jar before they're too big. Here's one way to use them. Let sliced cucumbers stand in cold limewater 24 hr. Wash in clear water 3 times, then soak in cold water 3 hr. Drain. Combine salt, celery seed, ground cloves, pickling spices, vinegar and 2¼ lb. sugar. Pour over cucumbers and let stand 12 hr. or overnight. Stir now and then while in limewater and, too, while in syrup. Add another 2¼ lb. sugar before ready to then boil all 40 min. and pour into sterilized jars while hot. Use knife to get air out of jar and seal. Do not make

(Continued on next page)

more than one batch of pickles in same limewater. Lime? Yes, like one gets at a lumber yard. Hydrated lime. Kind I have on hand is "Snowflake". You'll have to add water a little at a time and make a paste as lime is hard to dissolve but it does--or most of it after a while. Then add to 2 gal. water. Recipe needs 4½ lb. sugar all told. Add half first, otherwise pickles tend to shrink. Green food color can be added for a greener pickle if desired.

Mrs. George Maxson
North Loup, Nebraska

Lime Sweet Pickles

1 gal. sliced cucumbers
2 cups lime
2 gal. water
8 cups vinegar
9 cups sugar
1 tbsp. salt
1 tsp. mixed pickling spice
1 tsp. celery seed
1 tsp. whole allspice
Few sprigs dill

Use any size cucumbers (just so they aren't seedy). Dissolve lime in water, add cukes and let stand 24 hr. Stir occasionally after 24 hr., drain and discard limewater. Wash cucumbers through 3 waters. Return to jar and cover with fresh cold water. Let stand 3 hr. (No longer). Drain well. Make syrup of vinegar, sugar, salt and spices (tied in bag). Bring to boil. Pour over drained pickles. Let stand overnight. In morning, put pickles and syrup in large kettle and boil 35 min. Pack while hot in jars and seal. Add a few sprigs of dill to the top of your jars before sealing for variety.

H. F. Gerdes
Clay Center, Nebraska

Lime Sweet Pickles

7 lb. cucumbers
2 cups slaked lime
2 gal. cold water
2 qt. vinegar
9 cups sugar
2 tbsp. pickling salt
1 tsp. whole cloves
1 tsp. celery seed
1 tsp. pickling spice
Green food coloring (if desired)

Cut cucumbers in about ¼ in. slices and soak 24 hr. in limewater (make a paste of the lime and a little water first and it will mix up better). Stir occasionally--as lime will sink to bottom. Drain, wash off, soak in clear, cold water 12 hr. Boil (then cool) a syrup made of vinegar, sugar, pickling salt, cloves, celery seed and pickling spice. (Put cloves, celery seed and pickling spice in cheese-cloth bag). You may add green food coloring. Drain off clear water and put pickles in cooled syrup. Let stand 12 hr. Bring to boil and simmer 35 min. Pack and seal in hot jars. The time element is essential in these pickles so I have made up a schedule (so I wouldn't be working at 4 a.m. sometime). Here it is:

9 a.m.--Put cukes in limewater (24 hr.)
9 a.m.--Put cukes in clear water (12 hr.) Make syrup and cool.
9 p.m.--Put cukes in syrup (12 hr.)
9 a.m.--Cook cukes 35 min. Put in jars and seal.

Mrs. H. J. Hoffman
Minden, Nebraska

Lime Sweet Pickles

7 lb. cucumbers
2 gal. water
2 cups lime (garden variety)
2 qt. vinegar (white)
1 tbsp. salt
9 cups white sugar
1½ tsp. celery seed
1½ tsp. whole cloves
1½ tsp. mixed pickling spice
Green color if desired

Slice cucumbers as desired. Cover with cold water and lime. Let stand overnight. Drain, wash well, cover with cold water. Let stand 3 hrs. Drain well. Cover with the syrup (heated enough to dissolve sugar) of vinegar, sugar, celery seed, whole cloves, pickling spice and salt. A few drops of green food coloring can be added if desired. Next day put on stove and simmer 35 min. or until clear. Put in jars and seal.

Mrs. Lawrence Dokulil
Omaha, Nebraska

Stickle Pickles

7 lb. cucumbers
1 cup lime
1 gal. water
1 qt. vinegar (white)
5 cups sugar
1 tsp. celery seed
1 tsp. cloves
1 tbsp. salt

These are sticks--they use the ones you missed until it was too late. Don't use yellow tinged cucumbers. Wash and split cucumbers and remove seeds. Cut into narrow sticks. Soak several hours or overnight in water and lime mixture. Pour syrup of vinegar, sugar, celery seed, cloves and salt over cucumbers, leave several hours or overnight. Cook ½ hr. Can hot.

Geraldine Walker
Sidney, Nebraska

Sticklers

Cucumbers
1 cup lime
1 gal. water
1 qt. vinegar (white)
1 tsp. salt
1 tsp. celery seed
1 tsp. whole cloves
Red or green food color (if desired)

Slice ripe cucumbers ½ in. thick lengthwise and 2 in. long. Discard any yellow part. Soak overnight in a stone jar. Use 1 cup lime in 1 gal. cold water. Next morning wash thoroughly in cold water. Soak in cold water 3 hr. Drain thoroughly. Boil syrup of vinegar, sugar, salt, celery seed and cloves (celery seed and cloves in a sack) and pour over cucumbers after it gets cold. Put pickles back into jar first. Next day heat the syrup with the cucumbers in it very slowly 30 min or until they are clear. Can in sterilized pint jars. Can add small amount of red or green paste food color for various occasions.

Mrs. W. H. Howarth
Tecumseh, Nebraska

Stickles

7 lb. cucumbers
1 cup lime
1 gal. water
1 qt. vinegar (white)
5 cups sugar
1 tsp. celery seed
1 tsp. whole cloves
1 tbsp. salt

Use large still-green cukes. Do not use seeds. Cut lengthwise in 3 or more sticks. Soak sticks overnight in dehydrated lime water. Wash thoroughly in morning and soak 3 hours in fresh cold water. Make mixture of vinegar, sugar, celery seed, cloves and salt. Pour mixture over cucumbers and let stand overnight. Boil mixture and cucumbers ½ hr. the next morning. Seal in hot jars.

Mrs. Glenn Melson
Lincoln, Nebraska

Sweet Stickles

7 lb cucumbers
1 cup lime
1 gal. water
1 qt. vinegar (white)
5 cups sugar
1 tbsp. salt
1 tsp. celery seed
1 tsp. whole cloves

Peel cucumbers, slice lengthwise and remove all seeds and soft pulp. Cut into sections about 2 in. long. Cut sections into sticks about ¼ in. thick. Discard any part of rind that has turned yellow. Soak overnight in brine made of lime and water. Drain well. Make syrup of vinegar, sugar, salt, celery seed and whole cloves (the celery seed and cloves in a cheesecloth bag). Bring to boil cucumber sticks in this syrup ½ hr. and seal hot.

Mrs. Matt New
Elkhorn, Nebraska

MUSTARD PICKLES

Jennie's Mustard Pickles

1 medium head cauliflower
1 qt. small green tomatoes
3 green peppers
1 qt. sliced onions
1 qt. cucumber chunks
1½ cups sugar
½ cup dry mustard
3/4 cup flour
1 tbsp. turmeric
7 cups cider vinegar
7 cups water

Break cauliflower into small pieces Cut tomatoes in chunks, the cucumbers and peppers in ½-in. pieces. Slice onions thin. Mix all together with 1 cup coarse salt and 4 cups water. Let stand overnight. Then drain and cover with boiling water, let stand 10 min. and drain again. Meanwhile combine sugar, mustard, flour, turmeric, vinegar and water, bring to boil and cook until thick. Stir in cauliflower, tomatoes, peppers, onions and cucumbers and cook until just tender, stirring occasionally. Pack in hot sterilized jars and seal. Makes 8 or 9 pt.

Earl May Seed & Nursery
Shenandoah, Iowa

Mustard Pickles

Cucumbers
1 cup sugar
1 cup salt
1 cup dry mustard
1 gal. vinegar

Sift sugar, salt and mustard together. Mix in vinegar. Wash and dry cukes (use 3 to 3½ in. cukes) and put in above mixture. Put weight on top to keep cucumbers under vinegar. Cucumbers help to make more juice. Use 2-gal. jar. Tie a cloth and paper over jar and keep in cool place. Or they can be sealed. These are nice with ham sandwiches. Half of above amount can be made.

Wilma Nyffeler
Alliance, Nebraska

Mustard Pickles

Cucumbers
2 cups vinegar
1/8 cup salt
1 oz. mustard

Quarter dill-size cucumbers and pack in jars. Pour liquid (cold) made of vinegar, salt and mustard over cucumbers and seal. Work dry mustard into a paste with some vinegar. Add salt and rest of vinegar.

Mrs. H. L. Schwenker
Lincoln, Nebraska

Mustard Pickles

Cucumbers
1 tbsp. salt
A little mixed spices
1 tbsp. sugar
1 tbsp. mustard
Vinegar

Wash cucumbers and put them in 1 qt. jars as packed as possible. Put in salt, spice, sugar and mustard. Fill jars with cold raw vinegar. Seal. Shake jar to mix spices through jar.

Mrs. Charles Bors
Wilber, Nebraska

Mustard Pickles (Tested)

Small cucumbers
1 gal. cold vinegar
2 cups sugar
1 cup salt
1 cup dry mustard

Wash and wipe cucumbers dry. Mix vinegar and mustard, then add salt and sugar. Pour over cucumbers. This is enough mixture for 7 qt.

Mrs. Walter Jeary
Seward, Nebraska

REFRIGERATOR PICKLES

Easy Refrigerator Pickles

Cucumbers
3 medium-size onions
4 cups sugar
4 cups vinegar
½ cup pickling salt
1 1/3 tsp. turmeric
1 1/3 tsp. mustard seed
1 1/3 tsp. celery seed

Mix sugar, vinegar, pickling salt, turmeric, mustard seed and celery seed. Mix well, but don't cook it. Slice onions in bottom of 1 gal. glass jar. Add thinly sliced pickles. Pour on above mixture and refrigerate. Ready to eat in 2 days.

Miss Melva Risse
West Point, Nebraska

Excellent Refrigerator Pickles

Cucumbers
3 onions
4 cups sugar
4 cups vinegar
½ cup pickling salt
1 1/3 tsp turmeric
1 1/3 tsp. celery seed
1 1/3 tsp. mustard seed

Onions and cucumbers sliced thin. Mix sugar, vinegar and spices together. Do not heat--this syrup is cold. Wash and sterilize three large peanut butter jars. Slice 1 onion in each. Wash and slice enough cucumbers to fill jars. Stir syrup well and pour over cucumbers and onions. Screw on lids. Refrigerate at least 5 days before using. Keep in refrigerator. These will keep a year or longer if not used.

Mrs. Harold Kinkle
Grand Island, Nebraska

Excellent Refrigerator Pickles

Cucumbers
3 onions (sliced thin)
4 cups sugar
4 cups white vinegar
½ cup salt
1 1/3 tsp. turmeric
1 1/3 tsp. celery seed
1 1/3 tsp. mustard seed

Slice cucumbers thin, Mix sugar, vinegar and spices. Do not heat this syrup. It is cold. Wash and sterilize 3 large peanut butter jars, slice 1 onion into each, use enough cucumbers (washed and sliced) to fill jars (packing in a little). Stir syrup well and pour over cucumbers and onions. Screw on lids, refrigerate at least 5 days before using. Keep in refrigerator until ready to use.

Mrs. J. F. Janousek
Lincoln, Nebraska

Excellent Refrigerator Pickles

Washed cukes
3 onions (sliced thin)
4 cups sugar
4 cups cider vinegar

½ cup pickling salt
1 1/3 tsp. turmeric
1 1/3 tsp. celery seed
1 1/3 tsp. mustard seed

Mix sugar, vinegar, salt and spices in pitcher. Stir well. Do not heat. Wash and sterilize 3 large jars like 3-lb. peanut butter jars or whatever you have. Slice onion in bottom of each. Slice cukes evenly and fill jars. Pack down. Stir syrup again and pour over cukes. Place a wad of wax paper in top to hold pickles under liquid. Place a piece of wax paper over top of jar and screw on lid. Keep in refrigerator until ready to serve. (At least 5 days). I've kept some a whole year and they are wonderfully crisp and fresh tasting. Like bread and butter pickles, only better. They seem to improve with age--if you don't eat them up too fast.

Mrs. Thomas McLellar
Spencer, Iowa

Ice Water Pickles

Cucumbers
Celery
Onions
Ice

1 cup cider vinegar
1 cup sugar
½ cup salt

Cut cucumbers and celery in desired lengths and soak in large pan with plenty of ice, 3 hr. Fix small onions same way--in separate pan. Add vinegar, sugar and salt. Stir often until all is dissolved. Pour over pickles (well-packed in jars) and seal at once--all cold.

Mrs. Joe Schatz
North Platte, Nebraska

Refrigerator Pickles

Cucumbers
3 onions
4 cups sugar
4 cups vinegar

½ cup salt
1 1/3 tsp. turmeric
1 1/3 tsp. celery seed
1 1/3 tsp. mustard seed.

This recipe calls for large peanut butter jars but I have used other ones with a wide mouth. Mix sugar, vinegar, salt, turmeric celery seed and mustard seed. Slice onions and put in bottom of jars. Slice cucumbers in jars and pack firmly. Pour over above solution. Do not heat or boil. Set in refrigerator and let set 5 days before using. Must be refrigerated all the time. Will keep months.

Mrs. Charles Weber
Columbus, Nebraska

Refrigerator Pickles

Cucumbers
3 onions
4 cups sugar
4 cups vinegar

½ cup canning salt
1 1/3 tsp. turmeric
1 1/3 tsp. celery seed
1 1/3 tsp. mustard seed

In 3 qt. jars, slice an onion in each and enough cucumber slices to almost fill each jar. Mix syrup of sugar, vinegar, salt, turmeric,celery seed and mustard seed but do not boil it. Pour it over cucumbers and onions. Cover each jar with a lid but do not seal tightly. Keep in refrigerator until used up. Will be ready in 5 days. Keeps 9 mo. to a year.

Miss Helen Kucera
Hastings, Nebraska

Refrigerator Pickles

Cucumbers
Onions
4 cups vinegar
4 cups sugar

½ cup salt
1 1/3 tsp. turmeric (or 2 tsp.)
1 1/3 tsp. mustard seed (or 2 tsp.)
1 1/3 tsp. celery seed (or 2 tsp.)

Mix vinegar, sugar, salt (use regular table salt, not iodized), turmeric, mustard seed and celery seed. Mix well but do not heat. Use 3 large peanut butter jars and fill jars with sliced cucumbers and press them down before putting juice over them. Slice onions thinly in each jar of cukes, using 3 onions, one for each jar of pickles you have. Fill jars with vinegar mixture and place in refrigerator. Keeps indefinitely when kept cold. Use second day.

Mrs. Willard Garner
Lincoln, Nebraska

Refrigerator Pickles

Cucumbers
Onions
4 cups sugar
4 cups vinegar

½ cup salt
1 1/3 tsp. turmeric
1 1/3 tsp. celery seed
1 1/3 tsp. mustard seed

Wash and sterilize 3 large peanut butter jars. Slice 1 onion in each jar. Wash and slice enough cucumbers to fill jars. Mix sugar, vinegar, salt and spices. Do not heat. This syrup is poured over cucumbers cold. Screw on lids. Refrigerate at least 5 days before using. These keep up to 9 mo. in refrigerator. Recipe taken from an area paper.

Mrs. Richard Downing
North Platte, Nebraska

Refrigerator Pickles

Sliced cukes
3 onions (sliced thin)
4 cups sugar
4 cups vinegar
½ cup coarse or pickling salt
1 1/3 tsp. turmeric
1 1/3 tsp. celery seed
1 1/3 tsp. mustard seed

Mix vinegar and spices. Do not heat. Sterilize 3 large peanut butter jars. Fill jars with sliced cukes. Pour syrup over. Screw on lids. Keep in refrigerator until ready to use. Do not can. (Do not forget to put the onions in.)

Olga Wangerin
Kensington, Kansas

Refrigerator Pickles

Cucumbers
Onions
Vinegar
Sugar
Salt
Mustard seed
Celery seed
Turmeric

Slice 1 onion in sterilized jars. Slice cucumbers thin until jar is filled. Make brine of 4 cups vinegar, 4 cups sugar, ½ cup salt, 1 1/3 tsp. mustard seed, 1 1/3 tsp. celery seed and 1 tsp. turmeric. Mix well and pour over pickles. Put lid on and set in refrigerator. Let set several weeks before using.

Mrs. J. C. Hawley
Marysville, Kansas

RELISH

Chili Sauce

1½ cups white vinegar
2 tsp. whole cloves
1 tsp. broken stick cinnamon
1 tsp. celery seed
5½ lbs. (22 medium) tomatoes,
washed, peeled and quartered
1 cup granulated sugar
1 tbsp. chopped onion
½ tsp. cayenne pepper
1 tbsp. salt

Combine vinegar, cloves, stick cinnamon and celery seed. Set aside. Cook half of tomatoes with half of sugar, onion and pepper. Boil vigorously, stirring frequently, 30 min. Stir in remaining tomatoes and sugar, onion and pepper. Boil vigorously, stirring frequently, 30 min. Stir in remaining tomatoes and sugar. Boil vigorously again 30 min., stirring as necessary. Strain vinegar, discard spices. Add spiced vinegar and salt to boiling tomato mixture. Stir constantly and cook 15 min. longer or until consistency desired. Pour immediately into hot sterilized jars to within 1/8 in. from top of jars. Seal each jar before filling the next. Makes 2 pt. Make only this amount of chili sauce for a bright product. Store in a dark place or wrap each cooled jar in brown paper before storing.

Mrs. Val Kuska
Lincoln, Nebraska

Chowchow

1 pt. sliced cucumbers
1 pt. sliced onions
1 pt. chopped sweet peppers
1 pt. chopped cabbage
1 pt. chopped green tomatoes
1 pt. lima bean
1 pt. cut string beans
1 pt. sliced carrots
1 pt. sliced celery
1 qt. vinegar
2 cups water
4 cups sugar
2 tbsp. celery seed
4 tbsp. mustard seed
2 tbsp. turmeric

Soak cucumbers, onions, sweet peppers, cabbage and tomatoes in salt water overnight, using ½ cup salt to 2 qt. water. In morning drain well. Cook lima beans, string beans, carrots and celery until just tender but not soft. Drain. Bring vinegar and water to boil, add sugar and seeds. Stir well. Add all vegetables and boil 10 min., stirring frequently to avoid burning. Place in hot , sterilized jars and seal at once. Caulifflowerets are a nice addition to chowchow but be careful not to overcook it. Cooking it separately is a good idea.

Mrs. H. O. Werner
Lincoln, Nebraska

Coney Island Relish

5 cups ground cucumbers
3 cups onions (chopped)
2 hot red peppers
2 sweet red peppers
2 green peppers
3½ cups finely chopped celery
3/4 cup salt
4 cups white vinegar
3 cups sugar
2 tsp. mustard seed
2 tbsp. celery seed

Measure 5 cups ground cucumbers. Grind and add to cucumbers the onions, hot red peppers, sweet red peppers, green peppers and celery. Combine with salt and 6 cups water. Let stand overnight. I add ice cubes. Drain thoroughly. Heat vinegar, sugar, mustard seed and celery seed. Bring to boil. Add drained vegetables. Bring again to boiling point and let simmer 10 min. Put in sterilized jars. Ready to eat when cool.

Mrs. Frank J. Hobbs
Lincoln, Nebraska

Cranberry Chutney

1 lb. pkg. cranberries, washed and stemmed
2 cups sugar
2 large pears, pared, quartered, cored and diced
4 oz. candied lemon peel

Combine in large saucepan, heat, stirring constantly, to boiling. Cover, cook 5 min. or just until starts to flow. Lightly skim off foam with spoon. Simmer, uncovered, 15 min. Serve hot or chilled. You might like this.

Mrs. Neil Chandler
Imperial, Nebraska

Cucumber Relish

2 qt. sliced cucumbers
1 cup sliced onions
1 tbsp. salt
1 pt. vinegar
1/8 tsp. pepper
1 tsp. mustard seed
1 tsp. cinnamon
½ tsp. turmeric

This is a darned good cucumber relish handed to me 50 years ago by Mrs. A. A. Abel, then of St. Paul, Nebraska. I continue to use it. Put salt on cucumbers and onions and let stand 1 hr. Mix vinegar, pepper, mustard seed, cinnamon and turmeric and let come to boil. Put in cucumbers and onions that have been well drained in a colander. Let come to rolling boil. Fill jars and seal.

Mrs. Louis Gross
Fremont, Nebraska

Cucumber Relish

1 gal. large cucumbers
1 gal. green tomatoes
1 qt. ripe tomatoes
4 to 6 onions
3 green peppers
3 red peppers
1 tbsp. celery seed
1 tbsp. mustard seed
1 tsp. allspice
1 tsp. cloves
¼ tsp. pepper
1 cup salt
1 pt. vinegar
5 to 6 cups sugar

(Continued on next page)

Grind cucumbers (peel and remove seeds), green tomatoes, ripe tomatoes, onions, green peppers and red peppers. Put celery seed, mustard seed, allspice, cloves and pepper in bag. Put salt on top of green tomatoes and let stand 2 hr. Then squeeze out all the juice. Then grind. Mix with rest of ground ingredients. Pour off all juice and add vinegar and sugar. Cook 30 min. and can hot.

Mrs. Frank Hahn
Johnson, Nebraska

Cucumber Relish

4 qt. ground cucumbers
2 qt. ground onions
5 tbsp. salt
6 cups sugar
6 cups vinegar
1½ tsp. celerv seed
2 tsp. turmeric
3/4 tsp. ground cloves
2 tbsp. mustard seed

Peel large cucumbers; leave small ones unpeeled. Scrape out large seeds. Put 4 tbsp. salt over cukes and let stand 2 hr. Put 1 tbsp. salt over onions and let stand 2 hr. Drain well. Cook sugar, vinegar, celery seed, turmeric, cloves and mustard seed. Add drained vegetables and heat to boiling for 7 to 10 min. Seal hot.

Mrs. J. M. Millstead

Elsie's Relish

2½ cups ground onions
3 cups ground peppers and celery
6 cups ground green tomatoes
3/4 cup salt
5 cups sugar
1 tbsp. celery seed
2 tbsp. mustard seed
1 tsp. cinnamon
1 tsp. cloves
1 tsp. allspice
3½ cups vinegar

Grind onions, peppers and celery first and save ½ cup of thisjuice. Continue grinding green tomatoes and discard juice. Stir in salt and let stand overnight. In morning, drain and rinse. In large kettle combine sugar, celery seed, mustard seed, cinnamon, cloves, allspice, vinegar and the juice saved above. Stir well. Add the ground mixture and cook down to desired consistency, stirring frequently to avoid burning.

Elsie Werner
Lincoln, Nebraska

Garden End Relish

5 cups cucumbers
3 cups onions
1 large stalk celery
2 red peppers
2 green peppers
4 cups vinegar
4 cups sugar
2 tsp. mustard seed
2 tsp. celery seed
1 tsp. turmeric

Grind cucumbers, onions, celery, red peppers and green peppers. Let stand in mixture of vinegar, sugar, mustard seed, celery seed and turmeric. Bring to boil. Boil 10 min. Seal in sterilized jars.

Mrs. Carol G. Lutz
Arcadia, Nebraska

Garden End Relish

3 or 4 carrots
8 onions
2 heads of cabbage
9 green peppers
9 red sweet peppers
½ cup pickling salt
3 pt. vinegar
4 cups sugar
2 tbsp. mustard seed
2 tbsp. celery seed

Grind carrots, onions, cabbage, green peppers and red sweet peppers all together, add salt, let stand 2 hr. Drain. Add vinegar, sugar, mustard seed and celery seed. Do not cook. Pack into sterilized jars and seal.

Mrs. Harry Saum, Sr.
Grant, Nebraska

Garden End Relish (Budreaux Sauce)

6 qt. cabbage
3 qt. green tomatoes
5 red peppers (seeds removed)
6 green peppers (seeds removed)
6 large onions
3/4 cup salt
5 cups sugar
2 qt. mild vinegar
2 tbsp. mustard seed
2 tbsp. celery seed
2 tbsp. turmeric powder (if preferred)

Grind or chop cabbage, green tomatoes, red peppers, green peppers, and onions. Oldtimers always chopped vegetables. Mix (in large kettle) salt, sugar, vinegar, mustard seed, celery seed (I prefer 2 cups cut-up celery) and, if preferred the turmeric powder. Cook all until it comes to a boil (30 min. or so). Seal in jars. Makes 6 qt.

Mrs. J. L. Scroggin
Oak, Nebraska

Golden Pickles Relish

6 lb. ripe cucumbers
3 ripe sweet peppers
3 green sweet peppers
3 large onions
¼ cup salt
2 cups brown sugar
2 cups white sugar
2 cups vinegar
1 tbsp. celery seed
1 tbsp. mustard seed
1 tbsp. turmeric

Wash, peel and chop or grind cucumbers. Grind ripe sweet peppers, green sweet peppers, and onions. Mix cucumbers, ripe sweet, green sweet peppers, onions and salt all together and let stand overnight. Wash well in cold water--drain vinegar. Heat brown sugar, white sugar, vinegar, celery seed, mustard seed and turmeric until sugars are dissolved. Pour over cucumber mixture. Cook 20 min., stirring often. Seal in scalded jars.

Mrs. H. A. Looker
Omaha, Nebraska

Green Sweet Relish (9 pints)

1 qt. green tomatoes
1 qt. onions
12 green peppers
6 red peppers
6 large cucumbers
1/3 cup salt
6 cups sugar
1 qt. vinegar
2 tbsp. celery seed
1 tsp. turmeric
1 tbsp. mustard seed

Grind tomatoes, onions, green peppers, red peppers and cucumbers, add salt, let stand overnight. Drain, cover with cold water 2 hr. Drain. Add syrup made of sugar, vinegar, celery seed, turmeric and mustard seed. Simmer 20 min. Seal.

Mrs. W. L. Kinghorn
Lewiston, Nebraska

Green Tomato Relish

4 qt. chopped green tomatoes
3 chopped green peppers
6 large onions
½ cup salt
5 tbsp. celery seed
5 tbsp. mustard seed
4½ cups sugar
4 cups vinegar

Combine chopped green tomatoes, chopped green peppers, onions and salt, mixing well. Then drain 1 hr. Combine sugar, vinegar, celery seed and mustard seed with pickles mixture and simmer 45 min. Put in jars while hot and seal.

Mrs. George Howard
Wahoo, Nebraska

Hot Dog Relish

5 cups ground cucumbers
3 cups ground onions
3 cups chopped celery
2 hot red peppers, ground
2 sweet red or green peppers, ground
3/4 cup salt

1½ qt. water
1 qt. white vinegar
3 cups sugar
2 tsp. mustard seed
2 tbsp. celery seed

Combine vegetables. Add salt, water. Let stand overnight. Drain. Heat vinegar, sugar, mustard and celery seed to boiling. Cook slowly 10 min. Seal in hot, sterilized jars. Makes 5 pt.

Mrs. Ervin Paben, Jr.
Columbus, Nebraska

India Relish

2 lb. cucumbers
2 lb. green tomatoes
4½ tsp. salt
1 pt. finely cut celery or cabbage
1 cup ground onion
1½ cups ground green or red sweet peppers
2 tbsp. finely chopped hot red peppers

1½ cups sugar
2 cups cider vinegar
2¼ tsp. salt
¼ cup white mustard seed
2 tsp. celery seed
¼ tsp. turmeric
1/8 tsp. each of ground mace and cloves

Choose firm, green cucumbers about 6 in. long and 1¼ in. in diameter, and tomatoes that have the whitish color acquired just before ripening. Wash vegetables, remove stems, cores and blemishes. Put quartered tomatoes and cucumbers through food chopper, using coarse blade. Put into glass or enamel bowl, add salt, let stand overnight. Place in colander and press out and discard liquid; add next four vegetables, measures lightly, then add the remaining ingredients. Simmer 10 min. stirring occasionally. Pour into sterilized jars and seal with glass or enamel lined lids. Makes 4½ pt.

Mrs. Wallace C. Mundhenke
Milford, Nebraska

My Own Raw Relish

Cucumbers
Onions
Red peppers
Green peppers
Green tomatoes
Salt

Vinegar
Sugar
Mustard seed
Celery seed
Caraway seed

Soak about 12 cups coarsely ground garden vegetables (cucumbers, onions, red peppers, green peppers and green tomatoes) in ½ cup salt overnight. Rinse in cold water and squeeze dry. Cover with liquid made of 3 cups vinegar, 6 cups sugar, 3 tbsp. mustard seed, 1 tbsp. celery seed and 1 tbsp. caraway seed. Seal and store in cool place.

Mrs. Carol G. Lutz
Arcadia, Nebraska

Onion Relish

7 cups finely chopped onions
½ cup coarsley chopped green pepper
3/4 cup coarsely chopped red pepper
or 1-4 oz. can pimiento
¼ cup sugar
2 tsp. salt
2 tsp. ground mustard
1 tsp. prepared horseradish
2 cups vinegar (5% acid strength)

Prepare vegetables - let drain 15 min. before measuring. Mix all ingredients and simmer 10 min. Seal while hot. Yiled 2 to 3 pt.

Mrs. Carol G. Lutz
Arcadia, Nebraska

Pepper Relish

12 red peppers
12 green peppers
6 onions (optional)
4 cups sugar
2 cups white vinegar

Grind on meat grinder. Pour boiling water over mixture. Let stand 15 min. Squeeze out as much juice as possible. Repeat twice. Boil sugar and vinegar. Put mixture into syrup after you drain pulp out of hot water. Boil 15 min. Seal.

Mrs. Charles Bors
Wilber, Nebraska

Piccalilli

One peck green tomatoes
2 small heads cabbage
3 green peppers
4 large onions
6 large cucumbers
1 cup of salt
4 qt. vinegar
4 lb. sugar

Chop tomatoes very fine. Put tomatoes, cabbage, peppers and onions all in a stone jar, sprinkle salt over this and let stand overnight. In morning, drain and scald in 1 qt. vinegar and 2 qt. water. Take out with skimmer and drain in sieve or colander. Make syrup of 3 qt. vinegar and 4 lb. sugar. Let all boil together 30 min. Put in glass jars and seal or this will also keep in a stone jar.

Hattie Parady
Dalton, Nebraska

Piccalilli

2 cups vinegar
4 cups sugar, white or brown
2 tsp. allspice
1 tsp. cloves
½ pkg. cinnamon sticks
Salt
Peck green tomatoes
2 bunches celery
4 large green peppers
5 large cucumbers
5 large onions
3 red peppers (not hot)

(Continued on next page)

Mix vinegar, sugar, allspice, cloves and cinnamon sticks, bring to boil. Soak green tomatoes, celery, green peppers, cucumbers, onions and the 3 red peppers (use red for color and use more than 3 if you have them) in salt water overnight. Drain. Grind vegetables. Cook 20 min. in syrup. Seal hot.

Mrs. Clarence Meyer
Lincoln, Nebraska

Piccalilli

1 peck green tomatoes
2 small cabbage heads
Salt
1 gal. vinegar
Brown sugar
White sugar
Mixed pickling spice

Chop tomatoes and cabbage fine and put in salt water. Let stand overnight. Drain from brine. Mix vinegar and enough brown sugar to taste with a bag of mixed pickling spice. Put on to cook and when boiling hot put in tomatoes and cabbage. Stir well and boil 20 min. Use enough white sugar to suit your taste. Very good recipe (my mother's).

Mildred Saf
Oakland, Nebraska

Queen of Pickles

Cucumbers
Onions
3 cooking apples
3 hot red peppers
1 large cauliflower
Salt
8 cups brown sugar
1 qt. vinegar
¼ lb. mustard seed
2/3 cup flour
6 tbsp. dry mustard
1½ tbsp. turmeric

Chop coarsely enough cucumbers (peeled and seeded) and onions to make 1 qt. each. Combine in a large bowl with apples, chopped, and peppers, trimmed and finely chopped. Add 1 qt. small, white onions and 1 large cauliflower, trimmed and broken into pieces. Bring to a boil 4 qt. water and 2 cups salt. Pour brine over vegetables and let stand overnight. Drain vegetables thoroughly and put them in a large kettle. Mix together brown sugar, vinegar and mustard seed. Pour mixture over vegetables and bring to boil. Blend together flour, dry mustard and turmeric. Gradually stir into vegetable combination. Bring pickles to rapid boil, stirring constantly. Spoon hot mixture into hot, sterilized jars and seal securely.

Earl May Seed & Nursery
Shenandoah, Iowa

Raw Vegetable Relish

5 carrots
8 onions
6 red peppers
6 green peppers
2 heads cabbage
Salt
3 pt. vinegar
5 cups sugar
3 tbsp. celery seed
2 tbsp. mustard

(Continued on the next page)

Grind fine the carrots, onions, red peppers, green peppers and cabbage. Add ½ cup salt with water to cover. Soak 3 hr. or longer. Drain overnight. Dissolve vinegar, sugar, celery seed and mustard. Mix and seal cold. Do not cook.

Geraldine Walker
Sidney, Nebraska

Uncooked Relish

8 medium carrots
2 medium heads cabbage
8 red or green peppers
(or 4 of each)
12 medium onions

½ cup salt
3 pt. white vinegar
6 cups sugar
1 tsp. celery seed
1 tsp. mustard seed

Grind vegetables and add salt. Stir well and let set 2 hr. Put into colander and drain very well. Then put into container, stir in sugar, vinegar and spices which have been stirred together until sugar is dissolved. Do not cook. Store in refrigerator or pack into fruit jars and seal. Store in cool place. (Note: I use the scrappy leaves of cabbage after making kraut for this and have some 2 yr. old just as good as when I made it. Any combination of vegetables may be used.

Mrs. Mary Voborny
Newman Grove, Nebraska

RIPE PICKLES

Ripe Cucumber Pickles

Ripe yellow cucumbers
Vinegar
Mustard seed
Mixed spices

Peel cucumbers. Cut in halves lengthwise and scrape out seeds with spoon. Cut each half in fourths. Sprinkle with ½ cup white vinegar, 1 cup water, ½ tsp. mustard seed and 1 tsp. mixed spices. When boiling hard, add cucumbers. Let just come to boil over. Do not boil more. Pack in sterilized jars and seal.

Mrs. Louis Gross
Fremont, Nebraska

Ripe Cucumber Pickles

Cucumbers
Salt
Pickling spices
Vinegar
Sugar

Soak cucumbers 3 hr. in salt water after seeds have been taken out and peeled then drain. Add 1 cup vinegar, 1 cup sugar, 1 cup water and pickling spices. Put in cucumbers and cook until clear, can and seal. This makes 2 qt.

Mildred Saf
Oakland, Nebraska

Ripe Cucumber Pickles

Cucumbers
Alum
Sugar
Vinegar
Cloves
Cinnamon

Cut cucumbers in halves lengthwise. Cover with alum water, allowing 2 tsp. powdered alum to each qt. of water. Heat gradually to boil-point, then let stand on back of range 2 hr. Remove from alum water and chill in ice water. Make syrup by boiling 5 min. 2 lb. sugar, 1 pt. vinegar, with 2 tbsp. each whole cloves and stick cinnamon tied in piece of muslin. Add cucumbers and cook 10 min. Remove cucumbers to a stone jar and pour over syrup. Scald syrup three successive mornings and return to cucumbers.

Raymond J. Larson
Plattsmouth, Nebraska

Ripe Cucumber Pickles

12 ripe cucumbers
½ cup salt
1½ cups sugar
4 tbsp. mixed spices
4 cups vinegar
½ cup water

(Continued on next page)

Wash and dry yellow ripe cucumbers, cut in two and take out seeds. Cut outside rim of solid meat into desired sized pieces. Sprinkle with salt and let stand in a cool place about 18 hr. Rinse and drain. Add sugar, spices (in bag) ½ cup water to vinegar. Heat to boiling. Add cucumbers. Simmer 15 min. Pack hot cukes in hot jars. Heat syrup to boiling. Fill jars with syrup and seal at once.

Mrs. Thomas McLellar
Spencer, Iowa

Ripe Cucumber Slices

12 ripe cucumbers
½ cup salt
1½ cups sugar
1 tsp. whole cloves
1 tbsp. whole allspice
1 long stick cinnamon
1 tbsp. mustard seed
4 cups vinegar

Cut cucumbers in thick slices crosswise. Remove seeds. Sprinkle with ½ cup salt and let stand overnight. Rinse. Drop into syrup made by boiling sugar, spices and vinegar 5 min. Simmer until tender. Pack into hot jars and seal at once.

Mrs. Reid Richards
Orleans, Nebraska

Ripe Cucumber Sweet Pickles

8 large ripe cucumbers
1/8 stick cinnamon
1 oz. whole cloves
6 cups sugar
1½ pt. vinegar
Salt

Pare, quarter cucumbers. Boil 1 min. in salt water (2 tbsp. salt in 1 qt. water). Drain and simmer in clear water until tender. Boil sugar, vinegar and spices 5 min. then pour over cucumbers. Let stand overnight. Drain off syrup and boil 10 min. and pour over cucumbers again. Flavor improves if this process is repeated several times. Seal.

Mrs. William J. Murphy
O'Neill, Nebraska

Ripe Pickle (Old Fashioned)

Cucumbers
Salt
Vinegar
Sugar
Mustard seed
Small stick cinnamon

Peel ripe pickles, cut lengthwise, scrape out seeds. Put in salt water overnight. Next day drain. Boil pickles in clear water with 1½ cup vinegar added. Drain pickles. Put pickles in jars, pour syrup over and seal. (Syrup is made of 1 cup water, 3/4 cup vinegar, 1 cup sugar, 2 tsp. mustard seed and cinnamon).

Mrs. Engnert Jensen
Exeter, Nebraska

Sunshine Pickle

Yellow ripe cucumbers
Salt
1 qt. vinegar
3 cups sugar
3 tbsp. pickling spices
1 tbsp. turmeric

Use yellow ripe cucumbers for this recipe. Peel, core and cut into cubes (any size you like) and soak overnight in salt water. Next morning, cook in following syrup until done and seal while hot: vinegar, sugar, pickling spices and turmeric.

Mrs. Walter Jeary
Seward, Nebraska

SACCHARIN PICKLES

Saccharin Pickles (For Diabetic)

2 2/3 qt. vinegar
1 1/3 qt. water
2 drops cinnamon oil
½ cup salt
2 drops clove oil
2 tsp. saccharin

Mix above together and pour over pickles cold. Enough for about 9 qt.

Wilma Nyffeler
Alliance, Nebraska

Saccharin Pickles

1 gal. vinegar
1 cup sugar
1 cup salt
½ cup dry mustard
1 cup grated horseradish
1 large handful mixed spices
1 large tsp. saccharin

One half batch will cover about 10 pt. small pickles packed in jars with liquid poured over cold.

Mrs. Carol G. Lutz
Arcadia, Nebraska

Saccharin Pickles

1 peck small cucumbers
9 cups vinegar
1 cup salt
1/3 tsp. oil of cloves
1 dram saccharin

Wash and pack cucumbers in jars (about 9 jars). Mix all dry ingredients (about 65 ½-grain tablets of saccharin make 1 dram) and add to the vinegar. Stir well to dissolve. Divide in 9 equal parts and put one part in each jar. Fill any remaining space in jar with cold vinegar and seal. (No cooking). Let stand about 6 wk. to ripen.

Mrs. Louis Gross
Fremont, Nebraska

Sweet Pickles for Diabetics

Cucumbers
Salt
White vinegar
Sucaryl
Pickling spice
Dill

Make brine of ½ cup salt to 1 gal. water and let cucumbers stand about 6 hr., then wash and split each one. Brine: 2 cups water, 1¼ cups white vinegar, 2 tbsp. Sucaryl, 1 tsp. pickling spice, few sprigs of dill. Bring to slow boil until they turn color. Then pack in sterilized jars and seal.

Mrs. Louis Shavlik
Linwood, Nebraska

Sweet Saccharin Pickles

Cucumbers
Salt
Powdered saccharin
Vinegar
Mixed spices

In bottom of 2 qt. jar put 2 tbsp. salt and ½ tsp. powdered saccharin. Fill jars with cucumbers which have been washed and dried. Add 2 more tsp. salt and 2 tsp. mixed spices. Fill jar with cold vinegar. Seal. This requires 8-10 wk. to make the pickles. 1 oz. of saccharin will make 24 qt. of pickles.

Earl May Seed & Nursery Co.
Shenandoah, Iowa

SAURKRAUT

Kraut

1 qt. shredded cabbage
1 tbsp. salt
1 tsp. sugar

Pack cabbage, salt and sugar into jar. Fill jar with boiling water and seal jar.

Mrs. Nettie Lovejoy
Valentine, Nebraska

Kraut in Glass Jars

Cabbage
Salt

Cut cabbage, salt and tamp it down in crock till it forms its own juice. Then put it in jars filling them 3/4 full packed real tight. Fill the rest of jar with juice, if you have enough. If not, add warm water. Put cover on real loose as it will leak. Put jars in pan or saucer. The first 2 days are the worst. Be sure there is enough water to cover the kraut at all times. When sour enough add cold water enough to fill jars and tighten the caps as much as you can.

Mrs. Charles Bors
Wilber, Nebraska

Old Fashioned SauerKraut

Cabbage
Onions
Apples
Pickling salt
Dill seed

Wash, quarter, core and shred nice hard cabbage heads. While shredding cabbage, add 3 large onions and 4 to 6 large, clean, cored apples. Weigh off 20 lb. Add ½ lb. pickling salt and ¼ cup dill seed. Mix thoroughly and let stand until it wilts and lets off water. May take ½ hr. Have stone jar ready. Take large handful of cabbage and squeeze all the juice you can, then pack firmly into stone jar. When jar is nearly full, cover with white cloth and large plate to weigh it down. Place 2 qt. jar of water on top to keep it firmly in place. There should be a brine of salt and water always on top. Allow 3 to 4 wk. to sour in a warm place. When cured, it is a yellow-white color. Pack in clean, hot fruit jars. Have it covered with the brine, using 1 qt. water and 2 tbsp. pickling salt. Process 30 min. in boiling water bath. Or simmer 15 min. and pack in hot jars, having enough of brine to cover when simmering and in jars covering kraut.

Mrs. Stanley Cada
Schuyler, Nebraska

Old Fashioned 'Sourcrout' (Sauerkraut)

Cabbage
Coarse salt
Coriander

It is a good idea to have a special barrel for making sourcrout, one in which vinegar or wine has been stored. Slice very white and firm cabbages in fine shreds with a cabbage slicer. At the bottom of the small barrel (keg or stone jar), place a layer of coarse salt and alternately a layer of cabbage and salt, being careful to have one of salt on top. As each layer of cabbage and salt, must be pressed down by a large and heavy pestle and fresh layers are added as soon as the juice floats on the surface. When the barrel or jar is full it may be seasoned with a few grains of coriander. When the barrel is full, it must be put in a dry cellar, covered with a clean cloth, under a plank and on this heavy weights are placed. At the end of a few days it will begin to ferment during which time the pickle must be drained off and replaced by fresh, until the liquor becomes clean. This should be done every day. Renew the cloth and wash the cover and weight and put back on and let stand a month. By that time the sourcrout will be ready to use. Care must be taken to let the least possible air enter the sourcrout and to have the cover perfectly clean. Each time the barrel must be opened it must be properly closed again. These precautions must not be neglected. In making the brine, be sure and use coarse or barrel salt. Taken from the White House Cookbook-1887.

Frank Field
Shenandoah, Iowa 51601

Saurkraut, Jar Method

Cabbage
Coarse salt
Sugar

In chopping cabbage, neither have it too coarse or fine. Slices about the width of the thickness of a nickle are about the right size. Never bruise cabbage as it will make the kraut soft and mushy. Remove the outside leaves of a fresh, hard, head of sound cabbage. Quarter and cut out the core. Chop or cut fine. Pack in sterilized jars and cover with boiling rain water. Run a knife down in the jar to remove the air bubbles. Add 1 tsp. coarse salt and one of sugar to each quart. Add enough water now to fill the jars, as some will have been removed with the removal of the air bubbles. Seal tightly and take to cellar as soon as cool. Use 2-piece glass top jars as they do not rust and kraut keeps better and longer.

(Second Method)

Prepare cabbage as above and add 1 tsp. coarse salt and 1 tsp. sugar. Cover with rainwater that has been boiled and allowed to cool; put on the glass part of the jar but do not seal. Set aside until the kraut ferments which will be about 3 or 4 days. If there isn't enough liquid left on the cabbage after it has finished fermenting, cover with the following: 1 qt. water, add 1 tsp. coarse salt and 1 tsp. of sugar. Have the soft water boiled and cooled before adding the sugar and salt. Seal and set away in the basement or cave. Be sure to use fresh, sterilized, hot, glass lids and rubbers when you seal the kraut. It keeps better in a cool, dark, dry location.

Ona Marie Weldon
Shenandoah, Iowa

Polish Kraut

Cabbage
Vinegar
Sugar
Alum
Pickle salt
Onions (if desired)

To each quart of cut cabbage add 1 tbsp. vinegar, 2 tsp. sugar and ¼ tsp. alum. Add onion if desired. Boil 1 gal. water and 1 cup pickle salt. Pour over cabbage and seal. Sterilize jars 20 min. in boiling water before packing.

Mrs. E. C. Volkmer
Talmage, Nebraska

Sauer Kraut

Cabbage
Pickling salt

Cut cabbage in fine long shreds. Pack in sterilized qt. jars (not too tight or the jars will break when it ferments). To each qt. jar add 1½ tsp. of pickling salt, fill with hot water. Seal and set in dark place. Ready to use in 3 wk. This may spew and bubble out but will settle back and seal itself. Keeps well and is nice and crisp. If you wish to use ½ gal. jars, use 1 tbsp. of salt.

Wilma Nyffeler
Alliance, Nebraska

Sauer Kraut

Cabbage
Salt

Use good sound heads of mature cabbage.
1 lb. salt to 40 lb. cabbage or 3½ tsp. with 5 lbs. 1 lb. fills 1 pt. jar. Remove outside green and dirty leaves, quarter the head and shred the cabbage finely. Put 5 lb. cabbage and 2 oz. salt into large pan and mix with hands. Pack gently into crock with potato masher. Repeat until crock is nearly full. Cover with cloth, weight and plate. During the curing process, kraut needs daily attention. Remove scum as it forms and wash and scald the cloth often to keep it free from scum and mold. Fermentation will be completed from 12 to 15 days. As soon as the kraut is thoroughly cured pack into jars, adding enough of the fruit juice or a weak brine made by dissolving 2 tsp. salt to 1 qt. water to fill the jars to within ½ in. of top of jar. Put on cap, screwing the band tight. Process in hot water bath 15 min.

Ralph J. Thompson
Curtis, Nebraska

Sauerkraut in Stone Jar

Cabbage | Salt

To 5 lb. cabbage add 3½ tbsp. salt. Mix thoroughly. Pack firmly and evenly with a potato masher into a stone jar (or crock) that has been washed in soapy water, rinsed and scalded. Repeat Step 1 until jar is filled to within 5 in. from top. Press firmly (do not pound) with masher to extract enough juice to cover cabbage by the time jar is filled. Keep cabbage covered with juice. Cover with 2 or 3 layers of white, clean cloth, tucking edges down against inside of jar. On top, place a scalded, heavy plate that just fits inside the jar or a paraffined board. Weight it down with a fruit jar filled with water so that juice comes over plate. Fermentation will begin the day following the packing. It works faster at high temperatures and the kraut is more likely to spoil at a high temperature. The best quality product is made at room temperature. (70 deg. F). Give the kraut daily care. Remove film as it forms and wash and scald cover cloth as often as necessary to remove mold and film. When bubbling stops (2 or 3 wk.--or 4 wk. in cold weather), tap jar or crock gently. If no bubbles rise, fermentation has ended. Pack into clean quart jars to within 1 in. of the top. Cover with sauerkraut juice. If you need more juice, add a weak brine (1½ tbsp. salt to 1 qt. water). Set jars in pan of cold water. Water should extend to shoulder of jars. Bring water slowly to boil; remove jars. Wipe off jar rims. Adjust lids. Process in boiling water bath 30 min. Remove jars from canner and complete seals. Zinc lids do not rust as rapidly as other lids. If you eat sauerkraut often and will use your supply before winter ends, it will hold in the stone jar in a cold room (55 deg. lower).

Sour Kraut

50 lb. cabbage
1 cup pickling salt
½ cup sugar

Start out with a large stone jar. 10-12 gal and 50 lb. cabbage. Clean all dry and bad leaves off cabbage. Cuts heads in half and shred (not too fine) into large dishpan or canner kettle. Mix salt and sugar well and sprinkle a handful over the cabbage, stir. Taste so you don't get too much salt. Pack down with the palm of the hand until water comes up over cabbage in bottom of stone jar. Repeat until you have all the cabbage cut. If you need more salt-sugar mixture, mix a small portion the same as before. Use half as much sugar as salt. Pack real well and cover with large plate and weight down with heavy object to keep cabbage under the juice at all times. (I use glass jugs or another stone jar filled with water). This will bubble up and then the bubbles will stop, which is the indication the kraut is done. I start to use it in about 1 wk. Then it should be put into a cool place. Always keep covered with water. If you have to add some, salt it to taste. If you think it is just right, then pack into fruit jars and process 5 min. in hot water bath. I like to keep it raw in the refrigerator best or cold cave. One can keep it in the stone jar for quite a while. I have always had complete success with this method.

Mrs. Mary Voborny
Newman Grove, Nebraska

SWEET PICKLES

Adams Pickles

14 large cucumbers
1 qt. vinegar
8 cups sugar
½ pkg. pickling spice (2 tbsp.)
2 rounding tbsp. salt

Pour fresh boiling water over cucumbers each morning. Fifth morning, cut lengthwise 2 times - then in inch lengths. Bring vinegar, sugar, spice and salt to boil each morning 3 mornings. Pour over pickles. Next morning, bring syrup to boil, add pickles, bring to boil again. Put in jars and seal. Delicious.

Mrs. Wm. G. Sprick
Fremont, Nebraska

California Pickles

2 gal. small cucumbers
1/3 cup pickling spices
1 qt. vinegar
7 cups sugar

Soak cucumbers 2 wk. in brine that will float an egg. Weight cucumbers down. Drain and soak 24 hr. in clear water, then 24 hr. in alum water (2 tbsp. alum to 1 gal. water). Wash in clear water and drain. Then pour over vinegar, sugar and spices. Heat to boiling point and reheat every day 7 days. If pickles are large, split when taking from brine. Pierce whole cucumbers with fork before boiling.

Mrs. Carroll D. Marcellus
Stuart, Nebraska

Cassia Bud Pickles

75 2-to-3-in. cucumbers
4 qt. water
2 cups coarse-medium salt
1 tsp. powdered alum
6 cups vinegar
8 cups sugar
1 oz. celery seed
1 oz. cassia buds

Cut cucumbers lengthwise. Heat water and salt to boiling. Cool. Pour over cucumbers, cover, let stand 1 wk. Drain, cover with boiling water, let stand 24 hr. Drain, cover with boiling water, add alum, let stand 24 hr. Drain. Cover with hot syrup made with vinegar, 5 cups sugar, celery seed and cassia buds. For 3 successive days, drain off syrup; add 1 cup sugar; heat to boiling, pour over pickles. (Total 8 cups sugar). Third day, drain, pack in sterilized jars, pour over hot syrup. Seal. Makes 14 pt.

Mrs. Dorothy Davidson
Coulterville, California

Cecila Pickles

1 gal. pickles
1 gal. water
1 pt. salt
1 heaping tbsp. alum
3 cups vinegar
6 cups sugar
1¼ tbsp. pickle spice

Mix water and salt. Put pickles in brine 6 days. Then add alum in hot water. Pour over pickles. Enough water to cover. Leave on 2 days. Cut pickles. Make syrup of vinegar, 5 cups sugar and spice. Boil and pour over pickles 3 days. Fourth day, add another cup of sugar. Pour over and seal. Very, very crisp.

Mrs. Otto Schmidt
Arlington, Nebraska

Company Pickles

10 medium cucumbers
8 cups sugar
4 cups cider vinegar
2 tbsp. mixed pickling spices
5 tsp. pickling salt

Cover cucumbers with boiling water. Let stand until morning. Repeat 3 times. Fifth day, drain and slice cucumbers into ½ in. slices. Combine all other ingredients, bring to boil and pour over cut cucumbers. Let stand 2 days. Third day, pack pickles into hot sterile jars, bring syrup to boil, pour over pickles. Seal. Do not be afraid of amount of sugar; 8 cups is correct and may well be secret of success for the pickles. Exceptionally tasty.

Mrs. W. F. Decker
Omaha, Nebraska

Crisp Cucumber Rings

9 long slender cucumbers
5 onions - medium size
½ cup salt
2 cups water
2 cups vinegar
1½ cups sugar
2 tsp. mustard seed
2 tsp. celery seed
½ tsp. Tabasco

Slice cucumbers and onions. Combine salt with water and pour over them. Let stand 3 hr. Then drain. Combine vinegar, sugar and spices; stir over heat until sugar is dissolved. Add cucumbers and onions. Boil 5 min. Add Tabasco and pack at once into hot jars and seal. Makes 2 qt. and 1 pt.

Mrs. Alvin L. Reed
Lincoln, Nebraska

Crispest Sweet Pickles

14 dill-size cucumbers
1 qt. vinegar (no water)
8 cups sugar
1 tsp. celery seed
3 sticks cinnamon

Pour boiling water over cukes. Next 3 mornings drain and pour fresh boiling water over. Fifth morning reheat syrup and pour over. Eighth morning, simmer pickles gently to heat thoroughly and seal hot. Do not use any salt or they will not be crisp.

Mrs. A. W. Tatum
Riverdale, Nebraska

Cucumber Pickles

2 gal. pickles
1 pt. salt
1 tbsp. powdered alum
3 pt. vinegar
8 cups sugar
2 tsp. celery seed
½ box cinnamon bark
2 handsful horseradish

Wash, split pickles. Put in stone jar. Pour 1 gal. boiling water, 1 pt. salt. Let stand 1 wk. Drain. Pour 1 gal. boiling water over, let stand 24 hr. Drain. Pour 1 gal. water over 1 tbsp. powdered alum. Let stand 24 hr. Drain. This is supposed to be done twice. Boil down liquid, pour over for 3 mornings. Let stand few days before using. Can in quart jars.

Mildred Saf
Oakland, Nebraska

Cucumber Pickles

Cucumbers
Vinegar
Sugar
Mixed spices
Mustard seed
Onion

Soak cucumbers overnight in cold water. Make syrup of 2 cups cider vinegar, 1 cup sugar, 1 tsp. mixed spices, ½ tsp. mustard seed and 1 cup water. Bring to boil, then add cucumbers and boil about 3 min. until they lose green look. Pack pickles solidly into sterilized jars, add 1 tsp. salt to each qt., ½ onion sliced, cover with boiling syrup and seal. These cucumbers may be sliced. I like to slice them lengthwise, cut into fourths.

Mrs. John Blinde
Johnson, Nebraska

Cucumber Pickles

1 qt. cucumbers sliced thin
1 cup vinegar
3 onions
2 green peppers
1 cup sugar
1 tsp. celery seed
½ tsp. cinnamon

Put onions and cucumbers sliced thin in salt water overnight and drain next day. Cook in vinegar and sugar and spice mixture for 20 min. and can while hot.

Mildred Saf
Oakland, Nebraska

Cucumber Pickles (Wonderful)

Cucumbers
Pickling salt
Vinegar
Alum
Chopped stick cinnamon
Whole cloves
Mixed pickling spice
Sugar

Wash and soak cucumbers any size. Bring to boil and cool this brine: ¼ pickling salt to 3/4 parts water (such as 3 qt. water to 1 qt. salt). Include ½ cup vinegar in brine. Pour cooled brine over whole cucumbers and forget 2 wk. Then cut cucumbers lengthwise, cover with cold water and 2 tbsp. alum for 24 hr. Then rinse well in cold water. Now tie in a cloth 2 tbsp. chopped stick cinnamon, 1 tbsp. whole cloves and 1 tbsp. pickling spice. Bring to boil 1 qt. vinegar and 2 qt. sugar, adding spice bag. Pour over pickles and repeat with same solution 4 days. It will almost be a syrup on 5th day. Seal cold without spice bag.

Mildred Saf
Oakland, Nebraska

Cucumber Pickles - Sweet

1 gal. cucumbers
6 cups sugar
1 cup water
1 tbsp. mixed spices
1½ qt. vinegar

Wash and dry fresh cucumbers. Cover with brine (1 cup salt dissolved in 1 gal. cold water). Let stand 24 hr. Drain. Puncture each cucumber in 2 or 3 places with needle. Simmer (do not boil) ½ of sugar, the spices, vinegar and water 30 min. Add cucumbers. Simmer 15 min. Let stand 2 days. Drain off liquid. Pack pickles in hot Ball jars. Add remaining sugar to liquid. Boil 5 min. Pour, while hot, over pickles; seal at once.

Earl May Seed Co.
Shenandoah, Iowa

Dutch Sweet Pickles

3 gal. 2½-to-3-in. cucumbers
3 cups salt
2 tbsp. alum
8 cups sugar
2½ qt. vinegar
Celery seed

Halve cucumbers. Add salt. Pour boiling water over to barely cover. Let stand overnight. Drain. Add alum cover with boiling water. Let stand overnight. Drain. Cover again with boiling water. Let stand until cool. Pack pickles into sterilized jars. Make syrup of sugar, vinegar, 1 qt. water with celery seed. Boil. Pour over pickles and seal hot. Sliced pickles can also be made with this recipe. They are very good and do not shrink.

Mrs. Minnie Dawson
Oshkosh, Nebraska

Easy Sweet Pickles

1 tsp. alum
1½ tbsp. pickling spice
2 tbsp. coarse pickling salt
2 cups white vinegar
1 cup water
3 or 4 cups sugar

Mix above ingredients. Wash pickles of any size and pack in ½ gal. jars, cover with above cold brine mixture and seal. Let set at least 6 wk., then drain, wash pickles and slice or cut into chunks and add sugar. Let set 24 hr. before using.

Mrs. Homer C. Blakeman
Merna, Nebraska

Easy Sweet Pickles

Small pickles
1 tsp. alum
1 tbsp. salt
1½ cup sugar
1½ tsp. pickling spices
1 cup water
Cold vinegar

Wash pickles (fine to use when you have only a few pickles at a time), place in sterilized quart jars. Add alum, salt, fill with vinegar and seal. Let stand 6 wk. or longer. To fix--pour off vinegar and discard. Cut pickles in shape desired. Bring to boil sugar, water and pickling spices. Add pickles and bring to boiling point. Remove from fire and let cool. Let stand 12 hr. or longer before using. The longer the better.

Mrs. William J. Murphy
O'Neill, Nebraska

Favorite Cucumber Pickles

2 gal. medium-size cukes
2 cups salt (not iodized)
Alum size of walnut
1 gal. scalding water

8 cups sugar
¼ cup mixed pickling spices
2 qt. vinegar

Wash cukes, dissolve salt in scalding water, pour over cukes. Let stand 1 wk. with cukes held under brine. Pour off brine, rinse and cover with boiling water and let stand 24 hr. Drain off again and cover with boiling water to which alum has been added. Let stand 24 hr. Drain and cut larger cukes in half. Mix sugar, spices and vinegar and heat to scalding. Pour over cukes. Next 3 mornings heat to scalding each time. Third morning pack in jars and seal.

Dorothy Straw
Lincoln, Nebraska

Finger Pickles

40 small cucumbers
2 qt. waters
1 cup coarse salt
½ tsp. alum

3 cups vinegar
4 cups sugar
1 tsp. celery seed
1 tsp. cassia buds

Boil water, salt and alum, let cool. Pour brine over cucumbers (split) boiling water. Let stand 24 hr. Drain. Cover with hot syrup of vinegar, 2½ cups sugar and spices. For 3 days drain off syrup, add ½ cup sugar each time, 3rd day, pack in jars. (7 pts.)

Mrs. Willard Downing
Seward, Nebraska

Ginger Crispies

1 gal. cucumbers
1 cup and 1 tsp. salt
1 tbsp. alum
1 tbsp. ground ginger

½ box whole pickling spices
8 cups sugar
5½ cups vinegar
1 tbsp. celery seed

Slice cucumbers thin. Add 1 cup salt to 1 gal. water. Let cucumbers stay in brine 5 days. Sixth day, soak cucumbers in alum water (1 tbsp. alum to 1 gal. water). Seventh day, drain and rinse. Boil 20 min. in ginger water (1 tbsp. ground ginger to 1 gal. water). Drain and rinse. Tie pickling spices in cloth. Add sugar, vinegar, celery seed, 1 tsp. salt, 2 2/3 cups water. Boil 20 min. or until slices are clear. Pack in jars, cover with hot liquid. Very crisp.

Vernetta Malicky
Burwell, Nebraska

Gladys' Sweet Pickles

2 gal. cucumbers
1 cup pickling salt
Alum
1½ qt. cider vinegar
Handful pickling spice
12-14 cups sugar

Slice cucumbers ½ in. thick. Add pickling salt and enough boiling water to cover. Let stand 1 wk. Stir or shake several times a day. Then wash with cold water, drain and cover with boiling water to which 1 heaping tbsp. alum has been added for each gallon of pickles. Let stand 24 hr. Drain, wash, then cover with boiling hot syrup made of vinegar, spice and sugar. For 3 mornings, heat syrup and pour over pickles. Third morning, pack pickles in jars, cover with boiling syrup and seal. Do not skimp on sugar.

Mrs. Ed Day
Madrid, Nebraska

Good as Heinz Pickles

Cucumbers
4 cups vinegar
4 tbsp. salt
3 tbsp. mixed whole spices
2 tsp. alum
4 cups sugar

Wash cucumbers and pack in 2 qt. jars. To each jar add 2 cups vinegar, 2 tbsp. salt, 1½ tbsp. mixed whole spices and 1 tsp. alum. Fill each jar with cold water and seal. Let stand 2 or 3 wk. before opening. A few days before using, remove from jars and wash in cold water. Split each pickle or cut in chunks. Place in refrigerator. Makes its own juice. Stir often. Can be eaten in day or two or you may can them.

Mrs. Eliza Stephens
Elm Creek, Nebraska

Good Neighbor Pickles

6 qt. medium-size cucumbers
1 cup coarse salt
3 or 4 green peppers
12 small onions
1 cup pimiento
1 qt. vinegar
6 cups sugar
2 tsp. mustard seed
1 tsp. celery seed
1 tbsp. turmeric

Slice cucumbers. Soak in water with salt for 3 hours. Drain but do not rinse. Add green peppers, onions (sliced) and pimientos, if preferred. Bring to boil vinegar, sugar, mustard seed, celery seed and turmeric. When boiling, add cucumber mixture and let come to boil. Pack in jars and seal. Green color can be added, if desired. Very crisp and delicious.

Mrs. Ed Blum
Herman, Nebraska

Little Krispies

75 small cucumbers
1 cup salt
1 tbsp. powdered alum
5 cups sugar
1 tbsp. celery seed
1 stick cinnamon

Cover cucumbers with cold water, add salt. Let set 1 wk. Weight down the cover so all pickles will be held under water. At end of week, split each pickle, put back in crock. Cover with boiling water. For 3 mornings pour off water and add fresh boiling water. Put alum in on 2nd morning. Fourth morning, pour off water, leaving pickles in crock. Cover with boiling solution of vinegar, sugar, celery seed and cinnamon stick (cut in 4 pieces). Reheat solution 3 mornings and pour back over pickles. Have boiling hot each time. Fourth morning, pack pickles in sterilized jars and pour boiling hot solution over them. Seal.

Mrs. Ora Randall
Mitchell, Nebraska

Martha's Dishrag (Copper Sponge) Sweet Pickles

Small to medium-size pickles
Salt
Alum
8 cups sugar
1 qt. vinegar
1 tbsp. mixed spices
6 sticks cinnamon

Soak pickles in salt water strong enough to float an egg for about 7 days for last pickles added to jar. Then soak out in fresh water each day for 3 days. Drain. Split every pickle. Boil pickles in alum water ½ hour--2 tbsp. alum per 2 gal. water. In bottom of kettle place a new, washed copper pan scratcher. Drain and rinse pickles well in cold water and drain again. Heat vinegar, spices and 2½ pt. sugar in cloth bag, with cinnamon among pickles (should cover 1 gal. pickles). Pour over pickles reheated every day; for 3 days adding 1 cup sugar each day. Pack pickles in sterilized jars and cover with boiling liquid. Seal. (Rinse copper sponge dry and seal in small clean jar to use next time).

Mrs. Reid Richards
Orleans, Nebraska

Marvelous Sweet Pickles

1 gal. medium cucumbers
1 gal. water with 1 cup pickling salt
1 tbsp. alum
1 tbsp. ginger
8 cups sugar
5½ cups vinegar
2 2/3 cups water
1 tsp. celery seed
½ box pickle spices

(Continued on next page)

Wash and slice cucumbers. Place in brine made with 1 gal. water and 1 cup pickling salt. Soak cucumbers in brine 4 days. Stir each evening. Fifth evening, remove cucumbers from brine, wash and return to jar and cover with 1 gal. of water with 1 heaping tbsp. of alum. Let this stand over night. Next morning drain alum water from cucumbers. Put cucumbers in large kettle, add 1 gal. fresh water and 1 tbsp. of ginger. Boil 10 min. While pickles are boiling, make a syrup of the sugar, vinegar, 2 2/3 cups of water and celery seed and pickle spices which have been tied together in bag. Cook this syrup until it boils well and becomes clear. Drain ginger water from cucumbers and discard. Pour hot syrup over pickles. Continue boiling until pickles become transparent. Seal in hot sterile jars.

Lydia Schweppe
Murdock, Nebraska

Marvelous Sweet Pickles

1 gal. medium cucumbers
1 gal. cold water
1 cup pickling salt
1 tbsp. alum
1 tbsp. ginger
8 cups sugar
5½ cups vinegar
2 2/3 cups water
½ box pickling spice (tie in cloth)
1 tsp. celery seed

Wash and slice cukes rather thin. Place in brine made of 1 gal. water and 1 cup pickling salt. Soak cukes in brine 4 days. Stir each day. Fifth evening, remove cukes from brine, wash and return to jar and cover with 1 gal. water and 1 tbsp. alum. Mix well, let stand on cukes overnight. Next morning, drain alum water; now place cukes in a large kettle and add 1 gal. fresh water and 1 tbsp. ginger. Boil this 10 min. While this is boiling with cukes, prepare syrup from sugar, vinegar, pickling spice and 2 2/3 cups of water, with celery seed loose in syrup. Boil. At end of 10 min. drain ginger water and discard. Bring syrup to good boil. Pour hot syrup over pickles; continue boiling until pickles become transparent. Seal in hot, sterile jars.

Mrs. Emil Walla, Sr.
Linwood, Nebraska

Marvelous Sweet Pickles

1 gal. medium-size cucumbers
1 gal. water (rain water if available)
1 cup pickling salt
1 tbsp. alum
1 tbsp. ginger
8 cups white sugar
5½ cups white vinegar
Pickling spices
1 tsp. celery seed
Cinnamon buds (if desired)
1 or 2 drops green coloring

Soak cucumbers in water and pickling salt 4 days. Stir occasionally each day. Remove cucumbers 5th day from brine, wash, return to jar. Cover with 1 gal. water and alum. Let stand overnight. Next morning, drain alum water off. Rinse in

(Continued on next page)

clear water. Put cucumbers in large kettle, add 1 gal. fresh water and ginger. Boil 10 min. While pickles are boiling make syrup of sugar, vinegar and 2 2/3 cups of water. Put pickling spices, celery seed and (if desired) cinnamon buds--all tied up in a thin bag--in syrup. Drain cucumbers. Discard ginger water. Cook syrup until it clears. Pour hot syrup over cucumbers and boil until transparent--about 20 min. Add coloring. Seal them hot. Sometimes at first they feel limp but when they cool a week or two they become crisp and firm. Seal in sterilized jars; they are cleaner than open jars. Use larger-size rather than smaller-size cucumbers, before seeds develop, and cut in slices ¼ in. or thereabouts. Put in gallon glass jars vinegar comes in since crock jars aren't available. Stir with wooden ladle that doesn't cut them.

Mrs. Angeline Kroese
Panama, Nebraska

Mrs. Spence's California Sweet Pickles

Cucumbers	1 qt. vinegar
Alum	3½ lb. sugar
2 rounded tbsp. ginger	Mixed spices (to taste)

Soak cucumbers 2 wk. in cold brine strong enough to hold an egg up. Cucumbers may be added as picked. Cut each cucumber and soak in cold water until free from salt, which will take a whole day. Change water frequently. Soak 24 hr. in alum water, using 2 tbsp. alum in water to cover 2 gal. pickles. Then wash well and boil ½ hr. in strong ginger water (Use 2 rounded tbsp. ginger for amount of vinegar used in this recipe). Rinse, drain and dry pickles with cloth. Put in jar and scald every day for 7 days or until pickles become clear with syrup made of sugar, vinegar, spices and 1 pt. water. Seal boiling hot. (Scalding may seem like a lot of work but the pickles are delicious. Drain the mixture from the pickles each day and bring it to boil and repour it over the pickles each day for 7 days). Use a small can of powdered ginger for a bushel of cucumbers. If you use warm ginger water to start your pickles it won't take quite so long. Then, when they begin to boil you can see how they change color, from the olive or yellow or brown that they have become in the brine. Boil the pickles until you think they are all as green as you think they will get, which is 30 min. or maybe a little more. Keep stirring and ducking them all under. They must all be cut or the small ones pricked with a fork.

Hilda Buettner
Madison, Nebraska

"One-Quart Pickles" (Or Two Pints)

Cucumbers	1 cup vinegar
1 medium onion	1 tbsp. salt
1 red or green pepper	Pickling spices (if desired)
1½ cups sugar	

Slice cucumbers until sterilized jar half full, then slice onion and pepper. Combine sugar, vinegar, salt and (if desired) a small amount of pickling spices in saucepan. Bring to rolling boil and pour over contents of jar. Seal and let stand at least 3 days before using.

Mrs. George Maxson
North Loup, Nebraska

"One-Quart Pickles" (Or Two Pints)

Cucumbers
1 medium onion
1 red or green pepper
1½ cups sugar
1 cup vinegar
1 tbsp. salt
½ tsp. mixed spice (if you wish)

Slice cucumbers until sterilized jar half full, then slice onion and pepper. Bring to full boil sugar, vinegar, salt and (if you wish) mixed spice. Pour over contents in jar and seal.

Mrs. Alvin L. Reed
Lincoln, Nebraska

Pickles

1 gal. medium-size cukes
1 gal. vinegar
1 cup salt
1 cup dry mustard
2 cups sugar
1 tsp. alum
Horseradish
Grape leaves

Wash cukes, pack in crock. Mix vinegar, salt, dry mustard, sugar and alum and pour over cukes. Stir every day for 4 or 5 days, then cover with horseradish or grape leaves and let stand 5 or 6 wk. until cured. If sealing in jars for winter, heat vinegar so jars will seal.

Mrs. Leon Tiahrt
Rapid City, So. Dakota

Pickles

Cucumbers
Salt
Alum
5 pt. vinegar
9 cups sugar
½ oz. celery seed

Soak cucumbers 8 days in brine (1 cup salt to 1 gal. water). Drain. Then soak pickles 54 hr. in alum water (1½ tbsp. alum to 1 gal. water). Scald and soak 24 hr. in mixture of vinegar, celery seed and 6 cups sugar. Reheat and pour over pickles 3 mornings, adding 1 cup sugar each morning. Split each pickle before putting in vinegar.

Mabel Coleman
Merna, Nebraska

Pickles

2 gal. large green cucumbers
2 cups pickling salt
8½ qt. water
7 cups vinegar
12 cups sugar
1 tbsp. alum
1 tsp. green or red food color
4 sticks cinnamon

(Continued on next page)

Here is an unusual recipe which also uses up the large green cucumbers. Make them red or green for holidays. Cut unpeeled cucumbers in thirds crosswise. Remove seeds. Slice in ½ in. rings. Add salt and 8 qt. water. Let stand 5 days, drain. Combine 1 cup vinegar, alum and food color in kettle. Add cucumber rings and fresh water to cover. Simmer 2 hr. Or divide rings in 2 parts. Simmer 1 with ½ tsp. red color, the other with ½ tsp green color. Drain, make new syrup by bringing to boil 6 cups vinegar, 2 cups water, the sugar and cinnamon. Pour over rings. Let stand overnight. Drain, make new syrup by bringing to boil 6 cups vinegar, 2 cups water, the sugar and cinnamon. Pour over rings. Let stand overnight. Drain, reheat syrup and pour over rings. Repeat 3 days. On 3rd day pack rings in hot sterilized jars. Pour on boiling syrup. Seal.

Mrs. Louis Gross
Fremont, Nebraska

Pickle Recipe

2 gal. medium-size pickles
2 cups salt
Alum (piece size of hulled Walnut)
8 cups sugar
2 qt. vinegar
¼ cup mixed spices
2 oz. cinnamon
1 tbsp. celery seed

Wash pickles. Dissolve salt in 1 gal. of scalding water. Pour over cucumbers. Let stand 6 days. Make sure all cucumbers are held under brine; put a weight on to hold them under. Pour off brine and rinse. Cover with fresh boiling water. Cover with fresh boiling water. Let stand 24 hr. Drain again and cover with boiling water to which alum piece has been added. Let stand 24 hr. Then drain and split each cucumber. Pack in jars. Make syrup of sugar, spices, cinnamon and celery seed. Pour over hot cucumbers and seal.

Mrs. William Tachovsky
Wilber, Nebraska

Quart Pickles

Cucumbers
1 small onion
1 pepper
1½ cups sugar
1 cup vinegar
1 tbsp. salt (scant)
½ tsp. mixed pickling spices (if desired)

Slice cucumbers until quart jar is about half full. Into this slice onion and pepper (green or red). Rings would be pretty. Pack these loosely. Combine sugar, vinegar, salt and (if desired) spices. Boil hard. Pour over contents in jar and seal. Do not use for about 1 wk.

Mrs. Ethel Borgman
Lincoln, Nebraska

Salt-Free Sweet Pickles

7 to 10 lb. cucumbers
2 cups air-slacked lime
2 gal. cold water
8 cups sugar
2 qt. vinegar
1 oz. whole allspice

Wash cucumbers and slice crosswise ¼ or 3/8 in. thick. Mix lime and water in large container. Add cucumbers. Soak 24 hr., stirring occasionally. Rinse at least 3 times in cold fresh water. Then soak 4 hr. in fresh cold water. Make syrup of sugar, vinegar and allspice. Heat to boiling point. Pour on well-drained cucumbers. Let set overnight. Next day simmer this 35 min. Pack in jars and seal.

Miss Helen Kucera
Hastings, Nebraska

Simple Sweet Pickles

1 gal. cucumbers
3/4 cup pickling salt
3½ cups vinegar
7½ cups water
1 tbsp. alum
1 tbsp. turmeric
6 cups sugar
1½ tsp. dill seed
1 tsp. pickling spices

Slice cucumbers. Cover with boiling water to which pickling salt has been added. Let stand overnight. Wash and drain. Simmer cucumbers 30 min. in mixture of 2 cups vinegar, 6 cups water, 1 tbsp. alum and 1 tbsp. turmeric. Rinse and drain. Pack cucumbers in jars. Combine 1½ cups vinegar, 1½ cups water, sugar, dill seed and pickling spices (tied in cloth) and bring to rolling boil. Pour over pickles in jars and seal.

Lydia Schweppe
Murdock, Nebraska

Store Pickles

7 lb. cucumbers
Salt
2 cups vinegar
Green grape leaves (or horseradish leaves)
1 tsp. alum
6 cups vinegar
6 cups sugar
1 oz. allspice
1 oz. stick cinnamon

Use cukes about 3 in. long. Place in stone jar and cover with brine (1 pt. salt to 1 gal. water). Let stand 4 days, pour off, add cold water. Let stand 3 days. Wash and split cukes, put in vinegar, enough water to cover and 1 hand green grape leaves or horseradish leaves. Add alum. Simmer 2 hours (be sure not to boil). Pour off liquid and place in stone jar. Heat vinegar, overnight. Next morning bring liquid to boil and pour over pickles. Third morning, pack pickles in jars, add vinegar mixture (boiling hot) and seal.

Mrs. Eliza Stephens
Elm Creek, Nebraska

Sweet Alum Pickles

Cucumbers
Salt
Powdered alum
2 qt. vinegar
4 lb. sugar
Mixed spices

In stone jar, make brine of salt and water that will float an egg. Put washed cucumbers in brine for 1 wk. Drain, add fresh water for 3 days. Cut cucumbers in short pieces or in half and weigh. Place 5 lb. in kettle and cover with 1 oz. of powdered alum and 5 pt. of water. Simmer 2 hours and drain. Add 1 qt. vinegar and 1 qt. water. Simmer again 2 hr. Make syrup of 4 lb. sugar and 1 qt. vinegar. Boil until sugar is dissolved, add pickles and mixed spices and bring to boil. These do not have to be sealed but can be if packed and sealed while cold.

Mrs. Zelda Stuck
Auburn, Nebraska

Sweet Candy Pickles

10 (or more) medium cucumbers
2 tbsp. mixed pickle spice
4 cups vinegar
8 cups sugar
5 tsp. pickle salt
Green food coloring

Cover cukes (whole) with boiling water. Let stand overnight. Drain. Repeat next 3 mornings. Fifth day, cut in ½ in. pieces. Combine sugar, salt, spice and vinegar. Bring to boil. Pour over pickles. Let stand 2 days. Third day, bring to boil. Put in sterilized jars. Seal.

Mrs. Dick Nelson
Lincoln, Nebraska

Sweet Celery Seed Pickles

Cucumbers
½ cup salt
1 tsp. powdered alum
3 cups vinegar
3 cups sugar
1 tsp. celery seed
1 tbsp. pickling spice

Slice amount of cucumbers you wish to make up. Let stand in salt water, using ½ cup of salt to a gallon of water and 1 tsp. powdered alum for 2 to 3 hr. Drain and rinse with clear water. Make syrup of vinegar, sugar, celery seed, pickling spice and 3/4 cup water. Boil together for a minute, then add pickles and bring to a boil again and boil about 1 min. Put in hot sterilized jars and seal.

Mrs. Harold Kinkle
Grand Island, Nebraska

Sweet Cucumber Pickles

Cucumbers
1 small onion
½ cup sugar
3 cups vinegar
1/3 cup salt
1 cup water
1 tbsp. celery seed
1 tbsp. white mustard seed

Cook sugar, salt, vinegar and water and cool. Put cucumbers, onion (cut up), celery seed and mustard seed in vinegar mixture. Enough for 1 qt. Let stand awhile before using so they soak in good.

Mildred Saf
Oakland, Nebraska

Sweet Cucumber Pickles

75 medium-size pickles
1 pt. coarse salt
3 gal. water
1 tsp. alum
9 cups sugar
5 pt. vinegar
Small amount whole allspice
Stick cinnamon

Wash pickles. Mix salt and 1 gal. water in crock. Put pickles in brine and weigh down. Set in cool place 1 wk. Remove pickles from brine. Pour out brine. Put pickles back in crock. Bring 1 gal. water to boil with alum. Pour over pickles. Let stand 24 hr. Pour off, add 1 gal. clear boiling water. Let stand 24 hr. Take pickles out of brine and split. Boil 6 cups sugar, vinegar, allspice and cinnamon. Pour this mix over pickles. Let stand 24 hr. Next morning, pour liquid off and bring to boil. Pour over pickles. Do this 3 mornings, adding cup of sugar each morning. Then boil liquid 3 mornings without adding sugar. Can seal or leave in open crock.

Mrs. William J. Murphy
O'Neill, Nebraska

Sweet Cucumber Pickles

30 cucumbers 4 in. long
2 qt. vinegar
16 cups sugar
¼ cup mixed spices
¼ cup salt

Wash cukes, drain, put in crock. Cover with boiling water. Let stand 24 hr. Drain cover again with boiling water. Let stand 24 hr. Repeat this 4 days in all. Fifth day, drain, rinse, cut in slices, put back in crock. Combine vinegar, sugar, spices and salt and bring to boil. Pour immediately over pickles. Let stand 24 hr. Pour off liquid and reheat to boiling. Repeat this process 4 days. Ninth day, reheat and seal or leave in crock. This is my favorite.

Mrs. William J. Murphy
O'Neill, Nebraska

Sweet Pickles

Small or medium-size cucumbers
4 cups vinegar
2 cups sugar
4 tbsp. pickling salt
3 tbsp. whole mixed pickle spices
2 tsp. alum

Wash and split cucumbers. Pack in 2 qt. jars and add to each jar 2 cups vinegar, 2 tbsp. pickling salt, 1½ tbsp. whole mixed pickle spices and 1 tsp. alum. Fill jars with cold water and seal. Shake jars to dissolve salt in liquid. Must stand at least 2 to 3 wk. before opening. A few days before using, remove pickles from brine and wash all spices and liquid off. Put in crock or enamel pan and add 2 cups sugar and 1/3 cup water. Put in refrigerator and stir often to dissolve sugar and get over pickles. Store in refrigerator.

Dorothy Fries
Upland, Nebraska

Sweet Pickles

Cucumbers
Salt
Water
Alum
Sugar
Vinegar
Pickling spice

Make brine of 1 pt. salt to 1 gal. water, pour boiling hot over cucumbers and weigh down and leave 1 wk. Split cucumbers, make solution of 2 tbsp. powdered alum to each gallon of water, pour over cucumbers boiling hot and leave 24 hr. Repeat with fresh alum solution and leave another 24 hr.; do not weigh down cucumbers in alum solution. Make syrup of 1 cup sugar to each cup vinegar, add pickling spice tied in a bag and boil in syrup. Place cucumber in jars and pour boiling hot syrup over them, leave 24 hr. pour off syrup and add as much sugar as used before and pour over cucumbers boiling hot and seal.

Mrs. C. H. Paulsen
Aurora, Nebraska

Sweet Pickles

Cucumbers
Salt
Water
1 qt. vinegar
5 cups sugar
½ cup spices (cloves, stick cinnamon, mustard seed, celery seed)

Make brine of 1 part salt and 9 parts water. Cover fresh-washed cucumbers with brine. Let stand 1 wk. Drain. Cover with fresh cold water next 3 days, changing water each day. Drain. Cut in pieces--chunks--boil in fresh water 10 min. Drain. Cover with syrup (brought to boil) made of vinegar, sugar and spices. Following 3 mornings, drain syrup from pickles and heat to boiling and pour syrup back on pickles. Third morning, put in jars and seal.

Mrs. Clarence Meyer
Lincoln, Nebraska

Sweet Pickles

7½ lb. cucumbers
Dehydrated lime
Water
2 qt. white vinegar
4 lb. white sugar
1 tbsp. whole allspice
1 tbsp. celery seed
1 tbsp. mixed pickling spice

Slice cucumbers thin. Cover with 1 cup dehydrated lime to each gallon of water. Soak 24 hr. Drain. Cover with cold water. Let stand 4 hr. Wash, drain good, cover with vinegar. Put pickles, spices and sugar in large kettle and simmer 1 hr. Seal while hot in pint jars.

Mrs. Ivar Saf
Oakland, Nebraska

Sweet Pickles

Cucumbers
1 tsp. salt
1 tsp. alum
Cider vinegar
2 cups sugar
1 stick cinnamon
Few cloves
Water

Slice pickles about ¼ in. thick. Fill jar, add salt and alum. Cover with cold cider vinegar and seal. Will keep indefinitely. After 2 or 3 mo., when ready to use, bring sugar, cinnamon, cloves and 3/4 cup water to boil. While syrup is heating, drain off vinegar from pickles, cover with boiling water. Drain, add syrup. Very crisp and so good. These can be made up as you use them.

Sweet Pickles

Cucumbers
Salt
Alum
Water
½ small can powdered ginger
3 pt. vinegar
1 pt. sugar

Put pickles (2½ to 3½ in.) in brine (salt and water) strong enough to float a fresh egg 10 days. I have added cucumbers so that some have been in brine 8 to 12 days. Drain and soak in powdered alum solution (2 tbsp. in water enough to cover). Next day boil ginger. When it starts to boil, put in pickles (which have been split open) and boil 7 min. Remove from ginger water and put in sweet pickle vinegar (made of vinegar, sugar and 1 cup water) while boiling hot. Heat this sweet pickle each morning for 4 mornings. Fourth morning, put in jar and seal. You may add any spice you wish.

Mrs. Hattie Brown
Oxford, Nebraska

Sweet Pickles

Medium-size cucumbers
Water
Salt
1 cup vinegar
2 cups sugar
1 tsp. celery seed
1 tsp. mixed pickle spice (can use other spices if you desire)

(Continued on next page)

Pour boiling water over cucumbers 6 days, changing water each morning. Seventh day, add enough salt to water so it doesn't taste flat. Eighth day, cut cucumbers in chunks or split and pack in jars and pour boiling mixture (vinegar, sugar, celery seed and mixed pickle spice) on and seal. Can pour hot syrup over pickles and then reheat 2 or 3 mornings before putting in jars and sealing.

Dorothy Fries
Upland, Nebraska

Sweet Pickles

7 lb. cucumbers
Salt
Water
Grape leaves (or horseradish leaves)
1 tsp. alum
4 pt. vinegar
3 lb. sugar
1 oz. whole allspice
1 oz. stick cinnamon

Put cucumbers (about 3 in. long or smaller) in stone jar and cover with brine made of 1 pt. salt and 1 gal. cold water. Let stand 4 days, then pour off brine and cover with clear, cold water. Let stand 3 days. Wash and split, regardless how small. Put in kettle. Add 2 cups vinegar and enough water to cover. Add handful of grape or horseradish leaves and a slightly rounding teaspoon of alum. Simmer 2 hr. (do not boil). Pour off liquid and put pickles back in stone jar. Heat mixture of 3 pt. vinegar, sugar, allspice and cinnamon and pour over pickles. Let stand overnight. Next morning, pour off liquid and bring to boil and pour over pickles again. Third morning, pack pickles in jar, add vinegar mixture, boiling hot, and seal.

Mrs. Arnold Witt
Lincoln, Nebraska

Sweet Pickles

300 small cucumbers
2 qt. cold vinegar
1 ginger root
6 tbsp. dry mustard
6 cups sugar
(Also 6 tbsp. sugar)
2/3 cup salt
(Also 6 tbsp. salt)

Wash cucumbers and wipe dry. Sprinkle with salt, cover with boiling water. Let stand overnight. In morning remove from brine, wipe dry and lay in crock. Add mustard, salt, sugar and ginger to vinegar and pour this over pickles. Set in cool place. Each morning add ½ cup sugar until 6 cups have been used. Seal in sterilized jars.

Mrs. L. Sarson
Bassett, Nebraska

Sweet Pickles

7 lb. cucumbers
1 pt. salt
4 pt. vinegar
Grape leaves (or horseradish leaves)
1 tsp. alum
6 cups sugar
1 stick (1 oz.) cinnamon
1 oz. whole allspice
Water

Use cucumbers 3 in. long or smaller. Mix 1 gal. cold water and 1 pt. salt, pour over them and soak 3 days. Then soak 3 days in cold water. Split each pickle regardless of size. Put in kettle with 2 cups of vinegar and enough water to cover. Add a few grape or horseradish leaves, alum. Let simmer 1 hr. Drain. Mix sugar, 3 pt. vinegar, cinnamon and allspice, boil and pour over pickles. Next morning heat the same liquid and pour over pickles again. Third morning, put pickles in jars and pour hot liquid over them and seal. Do not put stick cinnamon in jars; it sometimes makes them dark.

Mrs. A. J. Hindman
Lincoln, Nebraska

Sweet Pickles

Cucumbers
½ cup pickling salt
2½ cups vinegar
2 cups sugar
6 whole cloves
10 allspice
1 cinnamon stick
Water

Wash cukes, put in jars. Make brine of 4 cups cold water, pickling salt, ½ cup vinegar. Bring to boil, pour over pickles and seal. When ready to use, drain off brine, cut each pickle lengthwise. Pour boiling water over, let stand 2 or 3 hr., drain and rinse with cool water. Cover with syrup made of 2 cups vinegar, sugar, cloves, allspice and cinnamon stick. Boil and pour over cukes. Let stand in open container 2 or 3 days until dark green.

Mrs. George J. Pool
Lincoln, Nebraska

Sweet Pickles

Cucumbers
Salt
Alum
2 tsp. ginger
1 qt. vinegar
7 cups sugar
Cinnamon sticks
Whole allspice
Water

For 2 gal. pickles. Make brine of 1 pt. salt to 1 gal. water. Put in brine 2 wk. Then wash off and cut in pieces. Keep in clear water 2 or 3 days. Change each day. Then put in alum water 1 day (use 2 tbsp. alum for 2 gal.) Then wash good and boil ½ hr. in ginger water (use 2 tbsp. ginger). Then dash in cold water. Wash and wipe with dry cloth. Put in jar and cover with hot syrup made of vinegar, sugar and 1 part water. Put cinnamon and allspice in cloth. Scald and pour on pickles. Heat 6 or 7 mornings. Seal while hot.

Esther Langhorst
Nickerson, Nebraska

Sweet Pickles

15 dill-size cucumbers
1 cup barrel salt
1 qt. vinegar
2½ lb. sugar (3 cups)
½ cup allspice

Use 15 cucumbers (1 cup salt to every 15 cucumbers). Put cucumbers and salt in water to cover them 1 wk. Then soak in cold water 3 days, changing water each morning. Slice in pieces about 1 in. thick and boil in clear water ½ hr. Drain. Put in stone jar. Cover with syrup made of vinegar, sugar and allspice (tied in cloth bag). Boil together until thick, pour over cucumbers, doing same for 3 mornings. Third morning, put pickles in jars, heat syrup and seal.

Mrs. C. A. Valder
Tekamah, Nebraska

Sweet Pickles

Cucumbers
Canning salt
2 cups sugar
1 cup vinegar
Mixed pickling spices
Stick of cinnamon
Water

Cut each end off good-size cucumbers. Then place cukes in stone jar. Add a little canning salt and pour boiling water over cukes 3 times. Use fresh water each time. (Do not let them stand in water too long or get soft. Put the first water on in the evening, then next morning and at noon). In evening cut cukes in slices or chunks and put back in stone jar. Make syrup of sugar and vinegar. Tie mixed pickling spices and stick of cinnamon in cloth. Boil and pour over pickles 3 times. Fourth day, put pickles in jars. Heat syrup, pour over pickles and seal. These are real good.

Mrs. Earl Ueberrhein
Ashland, Nebraska

Sweet Pickles

Any-size cucumbers
Salt
Alum
3 cups sugar
Vinegar
Cinnamon
Cloves
Water

Cover cucumbers with brine strong enough to float an egg. Let stand 12 days. Drain and soak in cold water 24 hr. Cut all lengthwise, cover with equal parts of vinegar and water, adding 1 tbsp. alum to each gallon solution. Bring to boil. Let stand overnight. Wash. Make syrup of sugar, 2 cups vinegar with cinnamon and cloves in bag. Make enough syrup to cover, bring to boil and pour over cucumbers 9 days. Seal. Takes time but worth it.

Mrs. Neil Chandler
Imperial, Nebraska

Sweet Pickles

Cucumbers
Salt
Alum
White vinegar
Sugar
Pickling spice
Water

(Continued on next page)

Heat 1 cup salt to 1 gal. water. Cover pickles. Cover with new brine 3 mornings. Split cukes and put in alum brine (1 tbsp. alum to 1 gal. water). Hot!!! Do this 2 mornings. For 4 gal. pickles, heat 1 gal. white vinegar, 16 cups sugar and 1 box mixed pickling spice. Pour over cukes and seal jars.

Olga Wangerin
Kensington, Kansas

Sweet Pickles

Cucumbers
Salt
4 cups vinegar
2 tsp. alum
4 cups sugar
1 tsp. mixed spices

Slice cucumber ½ in. thick. Cover with brine that will float an egg 3 days. Soak in cold water, changing every 3 days. Boil alum, 2 cups vinegar and 4 cups water. Let stand 2 days. Remove and throw away. Boil sugar, spices and 2 cups vinegar. Pour over pickles. bring liquid to boil 3 days. You may also use little cucumbers without slicing.

Mrs. H. L. Schwenker
Lincoln, Nebraska

Sweet Pickles

Cucumbers
Salt
2 cups vinegar
4 cups sugar
Small onions
Pepper slices
Dill
Mixed spices
Oil of cloves
Cinnamon

Clean and soak cucumbers in salt water 3 or 4 hr. Drain and wash off salt. Boil vinegar and sugar. Put cucumbers in while hot. Let stand about 10 min. Do not boil but keep hot. Seal. You can mix small onions, pepper slices if desired. Add dill to some, mixed spices to some. A drop or 2 of oil of cloves and cinnamon to change the flavor.

Mrs. Charles Bors
Wilber, Nebraska

Sweet Pickles

2 gal. sliced pickles
Salt
Alum
12 cups sugar
6 cups vinegar
2 tbsp. mixed spices
Water

(Continued on next page)

Put pickles in brine. Leave in 1 wk. Drain good. Put enough boiling water to cover with 1 tbsp. alum fresh each day for 3 days. Then cook syrup, made of spices, 6 cups vinegar and 6 cups sugar. Each morning for 3 mornings add 2 more cups sugar. Fourth day, strain juice and put pickles in jar and pour over hot. Seal.

Esther Langhorst
Nickerson, Nebraska

Sweet Pickles

50-75 cucumbers 4-5 in. long
Salt
1 tbsp. alum
5 pt. vinegar
10 cups sugar
½ oz. celery seed
½ oz. stick cinnamon
½ oz. cassia
Green food coloring
Water

Wash cucumbers. Cover with cold brine(1 pt. salt to 1 gal. water). Let stand 1 wk. Skim off scum as it forms. Drain and cover with boiling water. Add alum. Let stand 24 hr. Split in half and pour clean boiling water over. Let stand 24 hr. Remove to stone jar. Heat vinegar, 6 cups sugar, celery seed, cinnamon and cassia. Pour on pickles. Let stand 1 wk. After this each morning heat liquid to boiling point and add 1 cup sugar for 4 mornings. May add green food coloring. Eat.

Mrs. Edwin Trautman
Herrick, So. Dakota

Sweet Pickles

Cucumbers
Vinegar
Salt
Mixed pickling spices
Sugar
Water

Wash and pack whole small cucumbers in qt. jars. To each jar add 1 cup vinegar, 1 tbsp. mixed pickling spices. Fill jar with cold water and seal. After 3 wk. these will be ready to use. Wash off good, slice lengthwise and put in bowl. Add 1 cup sugar and 1½ tsp. water, cover. Leave in refrigerator 3 days and eat.

Mrs. J. E. Stratman
Wilcox, Nebraska

Sweet Pickles

Small-size cucumbers
Salt
Vinegar
Sugar
Whole pickling spice
Water

Use cucumbers 2½ to 3 in. long. Wash cucumbers and soak overnight in cold water with about 1 tbsp. salt. Next day--for about 1 qt. cucumbers use this measurement:

(Continued on next page)

Into a kettle add 1 1/3 cup vinegar, ½ cup water and 1 small cup sugar. Put ½ tsp. whole pickling spice into a small cloth bag and tie and put into kettle. Bring this all to boil, then add the drained cucumbers. Simmer 15 min. (Do not boil, just simmer). Pack hot pickles into jars and pour over liquid which you have brought back to boil. Seal jars. These are ready to eat in about 2 wk. You can throw the spice bag away or you can use it the next time you can pickles.

Mrs. D. W. Moench
Omaha, Nebraska

Sweet Pickles

1 gal. sliced or chunk cucumbers
3/4 cup pickling salt
½ pt. cider vinegar
1 tbsp. alum
1 tbsp. turmeric
Dill
1 pt. vinegar
4 cups sugar
Water

Cover cucumbers with water. Add pickling salt. Bring to boil, cool. Pour over pickles. Let stand overnight. Drain. Simmer ½ hr. in ½ pt. cider vinegar, 3 pt. water, 1 tbsp. alum and 1 tbsp. turmeric. Drain. Pack in jars. (I sometimes put dill in bottom). Heat 1 pt. vinegar, 4 cups sugar and 1 pt. water. Boil, pour over pickles. Seal.

Maude Garton
Ogallala, Nebraska

Sweet Pickles

1 gal. uniform-size cucumbers
Salt
1 tbsp. alum
3 cups vinegar
6 cups sugar
1½ tsp. mixed spices
Cloves
Cinnamon
Pickling spices
Water

Put cucumbers in salt water, 2 cups to a gal., let stand 1 wk. Drain well. Split or cut in chunks according to size and pour over them boiling water to which alum has been added. Let stand 24 hr. Drain well. Pour over them boiling water. Let stand 24 hr. Drain. Pour over hot syrup made of vinegar, spices and 5 cups sugar. Or, if you prefer, add some whole cloves stick cinnamon. (We prefer pickling spices). Boil the syrup and pour over the pickles each day for 3 days. Fourth day, add 1 cup sugar and boil. Pour over pickles and seal in sterilized jars.

Mrs. Paul Yeutter
Cozad, Nebraska

Sweet Pickles--Easy and Very Crisp

Cucumbers
2 tsp. alum
Salt
2 cups sugar
¼ tsp. turmeric
1 tsp. mixed pickling spice
1 cup vinegar
1 tsp. celery seed

(Continued on next page)

Pour boiling water over pickles for 6 days. Seventh day, cut to equal-size pieces and sprinkle alum (plus salt to taste) and cover with boiling water. Eighth day, drain, well-pack in jars. Make syrup of sugar, vinegar, turmeric, celery seed and spice. Boil and pour over pickles and seal. This syrup is enough for 1 qt. pickles; 4- or 5-in. pickles are best size for this recipe.

Miss Melva Risse
West Point, Nebraska

Sweet Pickles--No Cooking

Cucumbers
1 cup cold vinegar
1 tbsp. pickling spice
1 tbsp. pickling salt
½ tsp. alum
1 cup sugar

Chunk enough 3- or 4-in. cucumbers or smaller ones split, to fill jar. Mix and pour over the following: vinegar, pickling spice, pickling salt and alum. Complete filling jar with cold water. Seal and let stand 3 to 4 wk. To prepare for eating, drain well, put in bowl and cover with sugar. Let stand 24 hr. or more in refrigerator. Then eat. (They should be stirred frequently).

Mrs. Herbert Ehlers
Nebraska City, Nebraska

Sweet Pickle Sticks

9 medium cucumbers about 4½-in. long
6 cups boiling water
2 1/3 cups vinegar
2¼ cups sugar
2 tbsp. salt
1 tbsp. celery seed
3¼ tsp. turmeric (optional)
3/4 tsp. mustard seed

Select fresh, firm cucumbers. Wash and cut into sticks. Pour boiling water over. Let stand overnight. Next day, drain, pack solidly into clean jars. Combine remaining ingredients and boil 5 min. Pour this boiling hot liquid over the cukes in jars. Put cap on jar, screwing band tight. Process in boiling water bath 5 min. Makes 3 or more pt.

Mrs. Jack Kabella
Elsmere, Nebraska

Sweet Sliced Pickles

2 gal. cucumbers
14 small white onions
4 large green peppers
3/4 cup salt
10 cups sugar
1 tsp. celery seed
1 tbsp. turmeric
1 tsp. ground cloves
¼ cup mustard seed
2½ qt. vinegar

Slice cucumbers and onions. Cut peppers in fine bits. Mix salt through all. Weigh down. Let stand 3 hr. Drain well. Combine dry ingredients. Add vinegar, then vegetables. Place over low heat. Heat thoroughly but do not boil. Stir often. Seal. Makes a crisp pickle.

Mrs. S. C. Huffman
Lincoln, Nebraska

Unusual Sweet Pickles

Cucumbers
Salt
2 cups cider vinegar
2 tbsp. pickling salt
1½ tsp. mixed pickling spice
1 tsp. alum
3 cups granulated sugar

Sterilize 2 qt. fruit jars. Make brine (heating to boiling hot) of vinegar, pickling salt, pickling spice, and alum. Fill jars with fresh, washed cucumbers not too large. Pour boiling solution over cucumbers. Fill jars with boiling water and seal. Make only enough solution for one jar at a time. Process 3 wk. or several mo. Open jar, discard brine, rinse, split pickles, put in glass or enamel container. Sprinkle sugar and 3 tbsp. cold water over pickles. Put in refrigerator 24 hr. before using. Stir often to dissolve sugar. Keep refrigerated. Very different. Really good. Don't open before 3 wk.

Mrs. W. H. Howarth
Tecumseh, Nebraska

Sweet Pickles

Cucumbers (small or medium)
½ cup salt
1 pt. vinegar
3 lb. sugar
1 tsp. alum
1 stick cinnamon
1 tsp. mustard seed
½ tsp. whole cloves
½ tsp. whole allspice

Soak small or medium sized pickles in a solution of ½ cup salt to 1 gal. water mixed to make a brine. Soak for 3 days. Then to cold water change for 3 days. Drain, slice or chunk or both. On 7th day, cook until tender in weak (slowly) vinegar solution (1 qt. vinegar to 2 qts. water). Make syrup of 1 pt. vinegar, 3 lb. sugar, 1 tsp. alum, 1 stick cinnamon, 1 tsp. mustard seed, ½ tsp. whole cloves, ½ tsp. whole allspice. Pour hot syrup over pickles for 3 mornings and then seal hot. Heat syrup each of these three mornings. Delectable.

Mrs. L. Borgman
Lincoln, Nebraska

TIME PICKLES

Pickles Made Quickly

1 qt. dill pickles
2 3/4 cups white sugar
2 tbsp. pickle spice
½ cup tarragon vinegar

Slice and drain dill pickles. Let stand. Mix sugar, vinegar and spice. Let stand 4 hr. at room temperature. Put ½ amount of sliced pickles in a jar. Cover with ½ sugar mixture, then add remaining sliced pickles and sugar mixture. Cover with lid. Let stand in refrigerator 4 days.

Mrs. J. G. Porter
Lincoln, Nebraska

Quicky Pickles

Cucumbers
1 medium onion
1 pepper (red or green)
1½ cups sugar
½ cup vinegar
1 tbsp. salt

Wash and slice enough cucumbers to fill 1 qt. jar ½ full. The pepper can be less, the red is more for color. Mix sugar, vinegar and salt. Put some spices in bag. Then boil mixture and pour over pickles and seal. Takes 3 days to cure. This is nice to put in 2 pt. jars.

Mrs. Harvey Haecker
Pickrell, Nebraska

Five Minute Cucumber Pickles

Cucumbers (sliced or chunk) (enough for 4 qt.)
3 cups water
5 cups vinegar
5 cups sugar
1 tsp. black pepper
1 tsp. cloves
1 tsp. allspice
1 tsp. cinnamon
3 tsp. salt
1 onion (cut fine)

Mix vinegar, sugar, salt and spices (put spices in cloth bag). Bring to boil before putting cucumbers and onion in. Boil 5 min. They can be rather larger cucumbers so the seed wouldn't be hard, about dill size when done. Put in jars and seal. We like this recipe.

Mrs. Jessie Krofta
Table Rock, Nebraska

15 Minute Cucumber Pickles

Cucumbers (sliced or chunk) (enough for 4 qt.)
3 cups water
5 cups vinegar
5 cups sugar
1 tsp black pepper
1 tsp. cloves
1 tsp. cinnamon
3 tsp. salt
1 onion, medium size, cut fine

(Continued on next page)

Mix vinegar, sugar, water, salt and spices. Put on stove and let come to boil before putting in cucumbers and onion. Spices should be put in thin cloth sack. Be sure to leave room to swell while vinegar is coming to boil. Slice or chunk cucumbers and cut onion up fine. Prefer dill size or rather large cucumbers as these stay crisp better. Just so they are not seedy. Or any size can be used. When vinegar comes to boil, put in cucumbers and onions and let come to boil again. Boil 5 min. Then can in jars while hot.

Mrs. Harvey Wolfe
Lincoln, Nebraska

15 Minute Pickles

Cucumbers (sliced or chunked)
(enough for 4 qt.)
3 cups sugar
5 cups vinegar
1 tsp. pepper
3 cups water
1 tsp. cloves
1 tsp. allspice
3 tsp. barrel salt
1 tsp. cinnamon
1 onion, medium size, cut fine

Mix vinegar, water, sugar, salt and spices together and bring to hard boil. While vinegar mixture is coming to boil, slice or chunk cucumbers and onion. Add cucumbers and onion to boiling vinegar mixture and let boil 5 min. Seal hot. We like dill or large size cucumbers, although any size may be used.

Earl May Seed & Nursery Co.
Shenandoah, Iowa

24 Hour Dill Pickle

Cucumbers
Dill
Raw onion
1 bud garlic
¼ tsp. powdered alum
1 qt. vinegar
1 qt. water
½ cup pickling salt
1½ cups sugar

Cut dill size cucumbers in two lengthwise. Pack in qt. jar. Add dill, slice of raw onion, bud of garlic and powdered alum. Bring vinegar, water, salt and sugar to rolling boil. Pour over cucumbers at once and seal. Ready to eat in 24 hr.

Mrs. H. Winter
Pierce, Nebraska

3 Day Pickles

Cucumbers
Onion
Pepper (green or red)
1½ cups sugar
1 cup vinegar
Spices (if you like)

Sterilize a qt. fruit jar. Slice cucumbers and fill ½ full. Then slice on top onion and either a green or red pepper. Heat, do not boil, mixture of sugar and vinegar. You may add spices if you like. Ready to eat in 3 days. Makes 1 qt. of pickles.

Mrs. George Schmid
Monroe, Nebraska

5 Day Pickles

Pickles (1 gal. small)
½ cup salt
2 cups sugar
2 cups vinegar
Water

Each morning put on the pickles 2 qt. boiling water and ½ cup salt. Do this 3 days. Fourth morning, put on clear, hot water. Fifth morning, put on to boil the sugar, vinegar and 2 cups water. Put pickles in and boil 2 min. This makes pickles green color (spices).

Mrs. Harvey Wolfe
Lincoln, Nebraska

6 Day Pickles

Cucumbers (1 gal.)
1 pt. salt
Water
2 tbsp alum
2 pt. sugar
2 pt. vinegar
3 tbsp. mixed spice

To 1 gal. cucumbers add brine of 1 pt. salt and 1 gal. of water. Cover. Let stand 2 days. Drain, wash, split, cover. Let stand 2 days. Drain and simmer 30 min. using 2 tbsp. alum to 1 gal. water. Drain and rinse. Heat 2 pt. sugar, 2 pt. vinegar, 2 cups water and 3 tbsp. mixed spice. Pour over hot 3 mornings. Seal on 4th day.

Mildred Saf
Oakland, Nebraska

7 Day Sweet Pickles

Cucumbers (7 lb., size of man's thumb)
2 cups pickling salt
Water
6 cups vinegar
Grape leaves
1 tsp. powdered alum
6 cups sugar
1 oz. pickling spice
1 tbsp. celery seed
1 oz. stick cinnamon

Cover cucumbers with 2 cups pickling salt and 1 gal. cold water. Let stand 3 days. Drain and cover with cold water. Let stand 3 days. Drain and split every pickle. Place split pickles in large kettle or dishpan, add 2 cups vinegar and enough water to cover pickles, with a handful of grape leaves or horseradish leaves and 1 tsp. powdered alum. Simmer on stove 2 hr. Drain (throw this liquid and leaves away). Place pickles back in jar. Heat to boiling 4 cups vinegar, 2 cups water, 6 cups sugar--with 1 oz. pickling spice, 1 tbsp. celery seed, 1 oz. stick cinnamon tied in thin cloth (making 1 or 2 bags of them)so spices won't touch the pickles to make dark spots. Pour this liquid over the pickles and weight down so all will stay under. Repeat heating liquid and pour over (4 times in all). These will keep in this open jar or seal while hot. (Don't use a jar that has ever had lard in it--as pickles will get soft).

Mrs. Cyrus Tharp
Gothenburg, Nebraska

7 Day Sweet Pickles

Cucumbers (4-5 in. long)
1 cup salt (pickling)
Water
Alum
1 qt. cider vinegar
3 cups sugar
2 tbsp. mixed whole spice (put in cloth bag)

Use small pickles. Wash and pack in large container. Cover with salt water (1 cup pickling salt to 1 gal. water). Let stand 3 days. Drain, rinse and cover with cold water. Let stand 24 hr. Slice in thin slices and boil hard 30 min. in alum water (2 tbsp. to 1 gal. water). Drain and run cold water over. Then drop in ice and water (cool quickly). For syrup use vinegar, sugar and whole spices. Boil 5 min. Pour over pickles in the large jar. Next day drain and boil syrup again, adding 1 cup more sugar. Do same again 2nd day--1 cup sugar. Third day, pack pickles in jars you want to use, boil syrup again, remove spice bag, fill each jar with syrup and seal.

Mrs. Alice DeVries
Lincoln, Nebraska

8 Day Pickles

Cucumbers (7 lb.)
1 gal. water
8 green grape leaves
1 pt. salt
2 cups vinegar
1 tsp. alum

Wash cucumbers. Place in stone jar in brine made of salt and water. Let stand 4 days. Drain, then cover with clear fresh water each morning 3 consecutive days. Wash and split each pickle. Put in kettle with vinegar and water to cover pickles. Add grape leaves and alum and simmer 2 hr. Do not boil. Drain, place in stone jar. Make syrup of 3 pt. vinegar, 3 lb. sugar and 1 oz. whole allspice. Pour over pickles. Let stand overnight. Pour off liquid, reheat and pour over again. Third morning, pack pickles in hot jars. Heat same liquid, pour over pickles and seal.

Mrs. James Rader
Nebraska City, Nebraska

9 Day Pickles

Cucumbers (3-4 in. long) (market basket full)
6 pt. vinegar
12 cups sugar
2 oz. whole allspice

1. Place pickles in stone jar and cover with brine made of 2 pt. salt and 2 gal. cold water. 2. Let stand 4 days, then pour off brine and cover with clear cold water. Let stand 3 days. 3. Wash and split cucumbers regardless of size. 4. Put them in a kettle with 4 cups vinegar and enough water to cover. Add couple handfuls of grape leaves and 2 slightly rounding tsp. alum. Simmer 2 hr. 5. Pour off liquid and place pickles back in jar. 6. Heat 6 pt vinegar, 12 cups sugar, 2 oz. whole allspice. 7. Pour over pickles. Let stand overnight. 8. Next morning pour off liquid and bring to boil. 9. Pour boiling liquid over pickles again. 10. Third morning, pack pickles in jars, add vinegar mixture boiling hot and seal. 11. With 2 part lids may be necessary to process in hot water bath 10 min.

Mrs. Glen McCamley
Kearney, Nebraska

9 Day Sweet Pickles

Cucumbers (4 qt., small)	2½ pt. sugar
Water	1½ tsp. cloves
Salt	1½ tsp. cinnamon
Vinegar	

Use stone jar. Make brine of water and salt enough to hold up an egg. Put 4 qt. small cucumbers in jar, cover with brine. Leave 3 days. Drain and cover with water 3 days. Drain. Then boil 1 qt. vinegar and 2 qt. water. Let stand 3 days. Pour over pickles. Drain. Boil 1½ pt. vinegar, 2½ pt. sugar, 1½ tsp. cloves, 1½ tsp. cinnamon. Pour over pickles.

Mrs. Don Keetle
Ashland, Nebraska

9 Day Sweet Pickles

Cucumbers (medium and small)	1 qt. sugar
1 cup salt	1 qt. vinegar
Water	1 large stick cinnamon
1 tbsp. alum	1 tbsp. whole cloves

Select medium and small pickles. Wash, brush, put in stone jar. Cover with brine-- 1 cup salt to 1 gal. water. Let stand 1 wk. Take out, wash, split. Let stand in cold water overnight (add ice). Next morning simmer in alum water--1 tbsp. alum to 1 gal. water--30 min. Wash in clear water. Return pickles back to stone jar and cover with hot syrup of 1 qt. sugar, 1 qt. vinegar, 1 large stick cinnamon and 1 tbsp. whole cloves. Tie cinnamon and cloves in cloth. Boil, pour over pickles. Pour the hot juice over pickles after heated 8 mornings. On 9th can in jars. These stay nice and green and are so crisp.

Mrs. Frank J. Hobbs
Lincoln, Nebraska

12 Day Pickles

Cucumbers (75, 2-3 in.)	6 cups vinegar
Water	1 oz. celery seed
2 cups coarse salt	5 cups sugar
1 tsp. powdered alum	1 oz. cassia buds

Cut cucumbers lengthwise. Combine 4 qt. water and 2 cups coarse salt. Heat to boiling. Cool, pour over cucumbers, cover. Let stand 1 wk. Drain, cover with boiling water. Let stand 24 hr. Drain, cover with boiling water, add 1 tsp. powdered alum. Let stand 24 hr. Drain and cover with hot syrup of vinegar, celery seed, sugar, cassia buds. For 3 successive days drain off syrup and add 1 cup sugar. Heat to boiling and pour over pickles. Total amount of sugar is 8 cups. Third day, drain and pack into hot sterilized jars. Pour over hot syrup. Seal. Makes 14 pt.

Mrs. Albert H. Kuhr
Blair, Nebraska

12 Day Pickles

Cucumbers
2 cups salt
Water
1 tbsp. alum
1 qt. vinegar
2 cups white sugar
1 tbsp. cinnamon (or cassia buds)

To 1 gal. water add 2 cups salt. Cover pickles. Let stand 7 days. Seventh day, drain and cover with clear boiling water. Eighth day, drain and cover with boiling water to which you add 1 tbsp. alum to 1 gal. of water. Ninth day, drain and split ends of pickles and cover with syrup made by combining these ingredients: 1 qt. vinegar, 2 cups white sugar, 1 tbsp. celery seed, 1 tbsp. cinnamon (or cassia buds). Bring to boiling point. Tenth day, drain and add 1 cup sugar to each qt. of liquid. Heat and cover pickles. Repeat on 11th and 12th days except on 12th day pack pickles in pt. jars, pour solution over and seal.

Mrs. Ivar G. Saf
Oakland, Nebraska

12 Day Pickles

Cucumbers (75, 2-3 in.)
Water
2 cups coarse salt
1 tsp. alum
6 cups vinegar
5 cups sugar
1 oz. celery seed
1 oz. mixed spices

Cut cucumbers lengthwise, combine water and salt, heat to boiling. Cool. Pour over cucumbers, cover, let stand 1 wk. Drain, cover with boiling water. Let stand 24 hr. Drain, cover with boiling water, add powdered alum. Let stand 24 hr. Drain and cover with hot syrup of vinegar, sugar, celery seed and mixed spices. For 3 successive days drain off syrup and add 1 cup sugar, heat to boiling. Pour over pickles. Total amount of sugar is 8 cups. Third day, drain and pack into hot, sterilized jars. Pour over hot syrup. Makes 12 pt.

Mrs. Ervin Shirreffs
Tilden, Nebraska

12 Day Sweet Cucumber Pickles

Cucumbers
1 pt. salt
Water
1 tsp. powdered alum
1 pt. sugar
1 pt. vinegar
1 tsp. whole allspice

Leave whole to start. 1 pt. salt to 1 gal. boiling water, pour over pickles. Cover good. Let stand 7 days. Eighth day, drain and cover with 1 tsp. powdered alum to 1 gal. boiling water. Ninth day, drain and repeat fresh alum water. Repeat on 10th day. Eleventh day, cut each pickle longways. Put in jars, cover with boiling hot 1 pt. sugar to 1 pt. vinegar. Twelvth day, pour off vinegar solution and heat to boiling point and add another scant pt. of sugar to each pt. of vinegar, pour over pickles. Reheat vinegar again and add 1 tsp.whole allspice to each qt. pickles.

Mrs. L. E. Cassell
Grand Island, Nebraska

12 Day Sweet Pickles

Cucumbers
Salt
Water
1 tsp. alum

6 cups sugar
1 pt. vinegar
1 stick cinnamon
1 tbsp. mustard seed
½ tsp. whole allspice

Let pickles stand 3 days in brine of 1 cup salt to 1 gal. water. Drain and let stand in cold water 3 days. Seventh day, cut pickles in half and simmer in 3 cups water to 1 cup vinegar, 1 tsp. alum. Simmer 2 hr. (do not boil). Let stand 3 days in this, then drain and make syrup of 6 cups sugar, 1 pt. vinegar, 1 stick cinnamon, 1 tbsp. mustard seed, ½ tsp. whole cloves, ½ tsp. whole allspice. Put spices in bag. Pour hot over pickles. Do this 3 mornings. Seal on 3rd morning. I add ½ tsp. green coloring to syrup first morning it is made. This is a very good crisp pickle but be sure to simmer and not let come to boil. I use white vinegar.

Mrs. Everett E. Dunn
DeWitt, Nebraska

12 Day Sweet Pickles

Cucumbers
Salt
Water
6 cups sugar

1 pt. vinegar
1 stick cinnamon
1 tsp. mustard seed
½ tsp. whole allspice
½ tsp. whole cloves

Make a brine of 1 cup salt to 1 gal. water and let it stand 3 days on pickles. Drain and put clear water on 3 days. Drain and cook in weak vinegar water 20 min. and let them stand in this 3 days. Make syrup of sugar, vinegar, cinnamon, mustard seed, whole allspice and whole cloves. Pour this over pickles hot. Let stand 3 days. Reheat each morning. Third day, put in jars and seal.

Mrs. A. J. Hindmand
Lincoln, Nebraska

13 Day Cucumber Pickles

Cucumbers (2 gal.)
1 oz. alum
2 qt. vinegar
8 cups sugar
Water

2 tbsp. whole mixed spice
1 tbsp. celery seed
1 tbsp. whole spice
1 box stick cinnamon (1¼ oz.)

1st day-7th day--soak in salt solution, strong enough to float an egg. Every other day remove scum and stir pickles. Eighth day, drain and cover with boiling water. Ninth day, drain and cover with boiling water to which alum has been added. Tenth day, drain, add fresh boiling water. Let stand until cold. Drain, make syrup of vinegar, 2 cups sugar and spices (tied in a muslin bag). Heat and pour over pickles. Eleventh day, drain off syrup--add 2 more cups sugar. Heat and pour over pickles.

(Continued on next page)

Twelfth day, drain off syrup--add 2 more cups sugar. Heat and pour over pickles. Thirteenth day, drain off syrup and add remaining 2 cups sugar. Fill jars with pickles and cover with hot syrup. These may be kept in open jar. Note: chunk pickles may be made after the brining process.

Mrs. George Howard
Wahoo, Nebraska

13 Day Sweet Pickles

Cucumbers (2 gal. small)
Salt
Water
1 oz. alum

2 qt. vinegar
3/4 qt. water
8 cups sugar
Celery seed, mustard seed
or mixed spices

First day wash pickles and cover with salt solution which will float an egg. Second to 7th day, let pickles stand, removing scum and stirring every other day. Eighth day, drain, cover with boiling water. Ninth day, drain, cover with fresh boiling water to which alum has been added. Tenth day, drain, cover with boiling water and let stand until cold. Drain well, then cover with a hot syrup made of vinegar, water and spice and only 2 cups of sugar. Eleventh day drain off syrup, add 2 more cups sugar. Twelfth day, repeat as above, adding 2 more cups of sugar. Thirteenth day, add the last 2 cups of sugar. Pack pickles in jars, add hot syrup and seal.

Sisters of Precious Blood
Lincoln, Nebraska

14 Day Pickles

Cucumbers
Salt
Water
Alum (size of walnut)

8 cups sugar
2½ pt. vinegar
Pickling spices (in bag)

Split cucumbers and soak 1 wk. in brine (1 pt. salt to 1 gal. water). Drain off salt water, cover with boiling water. Let stand 24 hr. Drain, cover with boiling water in which alum (size of walnut) has been dissolved. Let stand 24 hr. Drain, cover with boiling water. Let stand 24 hr. Drain, cover with boiling syrup made of 8 cups sugar, 2½ pt. of vinegar and pickling spices in bag. Pour off syrup 4 mornings and reheat to boiling point and cover the cucumbers. Seal 4th morning. P. S. Must be split. Large ones can be ground after being put through the brine process. Use the spiced vinegar to make a relish.

14 Day Pickles

Cucumbers (2 gal. split)
1 pt. salt
1 tsp. powdered alum
Water

2½ qt. vinegar
10 cups sugar
Spices (if desired)

(Continued on next page)

Pour boiling salt water over pickles until covered. Let stand 1 wk. (weight down pickles). Pour off and put on boiling water. Let stand overnight, then drain. Pour off and put on boiling water. Let stand overnight, then drain. Pour on boiling water with 1 tsp. powdered alum in it. Let this stand overnight and drain dry again. Take 2½ qt. vinegar and 10 cups sugar and boil. Pour over pickles. Drain and boil 3 mornings straight and pour over pickles each morning. Add spices if you desire. Third morning, seal.

E. H. Williams
Norfolk, Nebraska

14 Day Sweet Pickles

Cucumbers (2 gal. whole or chunked)
2 cups salt
Water
1 tbsp. alum
3½ pt. vinegar
1½ pt. water
½ oz. celery seed
1 oz. cinnamon
6 cups sugar

Cover cucumbers with 2 cups salt and 1 gal. boiling water. Let stand 1 wk. Skim if necessary. Eighth day, drain and cover with 1 gal. boiling water. Ninth day, 1 gal. boiling water and 1 tbsp. alum. Tenth day, boiling water. Eleventh day, dry and cover with 3½ pt. vinegar, 1½ pt. water, ½ oz. celery seed, 1 oz. stick cinnamon and 6 cups sugar. Twelfth, 13th and 14th day, reheat vinegar, adding 1 cup sugar each day. Seal on 14th day. Remove stick cinnamon before canning.

Lester F. Whitney
Holdrege, Nebraska

14 Day Sweet Pickles

Cucumbers (75,3-4 in.)
2 cups salt
Water
1 tbsp. powdered alum
5 pt. strong vinegar
½ oz. celery seed
6 cups sugar
1 oz. stick cinnamon

Cut cucumbers into chunks or slice 3/4 in. thick. Should have 2 gal. after they are cut. Put into stone jar, pour over them 2 cups salt, 1 gal. boiling water. Let stand 1 wk., skim off every day if necessary. Eighth day, drain well and pour over them 1 gal. boiling water. Let stand 24 hr. Ninth day, drain again and pour over them 1 gal. boiling water and 1 tbsp. powdered alum. Let stand 24 hr. Tenth day, drain again, pour over them 1 gal. boiling water. Let stand 24 hr. Eleventh day, drain and put into a clean jar and pour over them the hot 5 pt. strong vinegar, ½ oz. celery seed, 6 cups sugar and 1 oz. stick cinnamon. Twelfth 13th, and 14th days, drain and reheat the liquid. Add 1 cup sugar each time. Seal in jars.

Mrs. Dale A. Rogers
Lincoln, Nebraska

14 Day Sweet Pickles

Cucumbers (2 gal.)
Salt
Water
1 tbsp. powdered alum

5 pt. vinegar
6 cups sugar
½ oz. celery seed
1 oz. cinnamon stick

Into a clean stone jar put 2 gal. cucumbers. Wash and cut in pieces (or chunks). Dissolve 2 cups salt in 1 gal. boiling water and pour while hot over pickles. Then cover and weight down pickles. Let stand 1 wk. Eighth day, drain, then pour 1 gal. boiling water over them. Let stand 24 hr. Ninth day, drain and pour 1 gal. boiling water with 1 tbsp. of powdered alum over pickles. Let stand 24 hr. Following day drain again, pour 1 gal. boiling water over them. Let stand 24 hr., then drain. Combine 5 pt. vinegar, boiling hot, 6 cups sugar, ½ oz. celery seed, 1 oz. cinnamon stick. Pour this over pickles, drain off 3 mornings, reheating it and adding 1 cup sugar each morning. With third and last heating pack pickles into sterilized jars, pour hot liquid over them and seal. This recipe is excellent. I've used it and love it.

Florence Benson
Omaha, Nebraska

14 Day Sweet Pickles

Cucumbers (dill size)
Salt
Water
Alum (size of walnut)

3 qt. sugar
1 qt. vinegar
4 sticks cinnamon
1 tbsp. whole cloves

Using brine of 1½ cups salt to 1 gal. of water, soak small to dill-size cucumbers 7 days. Drain. Pour fresh boiling water over pickles each day 3 or 4 days. Drain. Slice or chunk pickles. Use 1 gal. boiling water and alum size of walnut. Pour over pickles. Let stand until cold. Use 3 qt. sugar, 1 qt. vinegar, 4 sticks cinnamon, 1 tbsp. whole cloves. Boil and pour over pickles. Drain off syrup and boil each day for 3 days. Pack pickles in jars and pour boiling syrup over them. Seal.

Mrs. Edwin Cook
Trumball, Nebraska

14 Day Sweet Pickles

Cucumbers (2 gal.)
Salt
Water
1 tbsp. powdered alum

5 pt. vinegar
6 cups sugar
½ oz. celery seed
1 oz. cinnamon stick

Into a clean stone jar put 2 gal. cucumbers, washed and sliced lengthwise. Regardless of size of cucumbers must be sliced or they will shrivel. Dissolve 2 cups salt in 1 gal. boiling water and pour while hot over the pickles. Then cover and weight down pickles. Let stand 1 wk. Eighth day, drain. Then pour 1 gal. boiling water over them and let stand 24 hr. Ninth day, drain and pour 1 gal. boiling water with tbsp. powdered alum over the pickles. Let stand 24 hr. Following day or tenth day, drain again, pour 1 gal. boiling water over them. Let stand 24 hr. Then drain.

(Continued on next page)

For pickling mixture combine 5 pt. vinegar boiling hot, 6 cups sugar, ½ oz. celery seed and 1 oz. cinnamon stick. Pour this over pickles, drain off 3 mornings, reheating, pack pickles into sterilized jars, pour hot liquid over them and seal at once.

Mrs. Loren Case
Phillipsburg, Kansas

TOMATO PICKLES

Delicious Green Tomato Pickles

- 8 qt. tomatoes
- 12 onions
- ½ cup mustard seed
- ½ cup celery seed
- 1 tbsp. whole cloves
- 4 peppers (2 red, 2 green)
- 6 cups brown sugar
- 3 qt. vinegar
- Ginger root
- 5¢ anise seed

Tomatoes should be green and size of golf ball. Slice them 1/3 in. thick, sprinkle with salt and let stand overnight. Slice onions. Slice peppers in circles. Cook all ingredients for 3/4 hr. and seal jars.

Lincoln, Nebraska

Green Tomato Dills

- Many little green tomatoes
- Cucumbers
- Garlic
- Celery
- Green or red peppers
- Dill
- Vinegar
- Pickling salt

To each qt. jar add 1 clove garlic, 1 stalk celery, 1 green or red pepper, 1 head of dill and many little green tomatoes. Combine 2 qt. vinegar and 1 cup pickling salt. Cook this 5 min. Select small firm green tomatoes with or without stems. Pack cucumbers (sliced, halves, quarters, chunks, etc.), garlic, celery and tomatoes in sterilized jars and pour over to within a ½ in. of top of jar and seal. I use a stainless steel table knife to "press" out air in jars before sealing. Use 6 or so little tomatoes to a quart.

Mrs. George Maxson
North Loup, Nebraska

Green Tomato Pickles

- 1 peck green tomatoes
- 1/3 cup salt
- 3 qt. vinegar
- 5 cups sugar
- 1/3 cup whole mixed spices
- 2 qt. water

In evening slice, not too thin, the tomatoes, mixing the salt between layers. Let stand overnight. In morning drain well. Boil them in 2 qt. water and 1 qt. vinegar until color changes from green to white. Cook a few at a time and stir occasionally to insure even color. Drain well the second time and discard this vinegar. Dissolve 5 cups sugar in remaining 2 qt. vinegar. Add spices. Cook tomatoes in this vinegar 5 min. longer. Seal.

Mrs. William J. Murphy
O'Neill, Nebraska

Ripe Tomato Pickles

3 pt. tomatoes, peeled and chopped
1 cup shopped celery
4 tbsp. chopped red pepper
4 tbsp. chopped onion
4 tbsp. salt
6 tbsp. sugar
6 tbsp. mustard seed
½ tsp. clove
½ tsp. cinnamon
1 tsp. grated nutmeg
2 cups vinegar

Mix ingredients in order given. Put in stone jar and cover. This uncooked mixture must stand 1 wk. before using but may be kept a year.

Raymond J. Larson
Plattsmouth, Nebraska

Sliced Green Tomato Pickles

1 peck green tomatoes
6 onions
Coarse salt
1 qt. vinegar
4 lb. brown or white sugar
2 tbsp. mixed spices

Slice tomatoes (not too thin). Slice onions. Put in layers in stone, glass or enamel containers. Over each layer sprinkle coarse salt, ½ cup for all. Put weight on and leave overnight. In morning drain off salt water. Rinse with clear hot water, drain again. Now take vinegar, sugar and spices (in cloth), let come to good boil. Add onions and tomatoes, cook until just tender. Put in jars and seal.

Mrs. Carl Lamster
Crete, Nebraska

Spiced Green Tomatoes

Small green tomatoes
Salt
Whole cloves
4 lb. brown sugar
1 qt. white vinegar
2 or 3 sticks cinnamon

Use 6 qt. small green tomatoes about the size of crab apples. Place into 2 qt. boiling water with 2 tbsp. salt and boil just until skins crack. Remove from boiling water and plunge into cold water, then peel. Stick 3 or 4 whole cloves into each. Place into boiling syrup made of sugar, vinegar and cinnamon. Put into jars and seal at once.

Mrs. A. W. Tatum
Riverside, Nebraska

Tomato Sweet Pickles

2 gal. green tomatoes
1 tbsp. salt
1 gal. vinegar
1 qt. sugar
5 large onions (sliced)
2 tbsp. cloves
1 tsp. cinnamon

(Continued on next page)

Slice tomatoes and sprinkle with salt. Let stand overnight and drain next morning. Combine vinegar, sugar, onions, cloves and cinnamon, cook until tomatoes are done and seal. I always grind mine and we like it better. Is awfully good with any kind of meat.

Mrs. Fred Assarsson
Elsmere, Nebraska

Green Tomato Pickle

Tomatoes
Salt
Ground horseradish
Vinegar

Wash and slice as many tomatoes as you wish, sprinkle them all through with salt and let stand 24 hr. Then pour off water and drain thoroughly. Have a crockery jar ready (or other kind) and put in a layer of tomatoes and layer of ground horseradish until you have used up the tomatoes. Pour vinegar over this. These will be ready to eat in less than 1 wk.

Mabel Hovden
Mission Hill, So. Dakota

WATERMELON PICKLES

Baseball Pickles (Pickled Watermelon)

Watermelons
Dill tops
Grape leaves
Pure pickling salt
Vinegar

Gather watermelons when they are the size of baseballs or smaller. You can use the end of the garden melons. Scrub well and cut the stem end down to the end of the stem-joint so as to have an opening into the melon so the juice can replace the melons' own juice. Place in layers and cover each layer with dill tops and grape leaves. When the crock or large jar is full to within a couple of inches, cover with a plate and weight with a plastic bag full of pickling juice. This hold the melons under the liquid and they do not become soft. Make pickling liquid of soft water (1 gal. boiled 10 min. and cooled), pure pickling salt (1 3/4 cups) and vinegar (1 qt.). Mix with the water and add to the melons. After these have set 3 wk., they may be put in ½ gal. jars and sealed. Cut into slices when serving. This is a recipe given to me by a Hitterite lady in So. Dakota, and was very good. Some of the ladies used onions and garlic but these were boiled with the salt and the water and cooled as the onion or garlic tend to make the pickles soft (not always but some years it does). A 5-gal. jar will take about 2½ to 3 gal. of liquid for the pickles.

Mrs. Paul Harrison
Newport, Nebraska

Best Ever Watermelon Pickles

Watermelons
Salt
Vinegar
Sugar
Oil of cloves
Oil of cinnamon

Cut rind and nearly all the pink from the green of the melon and sprinkle lightly with salt. Let stand overnight. Pour off salt water. Cover with fresh water and simmer 10 or 15 min. Drain dry. Simmer again in liquid to proportions of 1 pt. vinegar, 4 pt. sugar, oil of cloves and oil of cinnamon. Boil 10 min. and seal.

Mrs. Carol G. Lutz
Arcadia, Nebraska

Delicious Watermelon Pickles

Watermelon rind
Sugar
White vinegar
Oil of cloves
Oil of cinnamon

Select thick rind of watermelon. Wash and cut into small pieces. Parboil rind in water just until it can be pierced. Be very careful not to get rind too soft. (Cut away any pink part of rind as you are preparing watermelon rind). For 7 lb. of cut-up rind use 3½ lb. sugar (7 cups). 1 pt. white vinegar (2 cups) ½ tsp. oil of cloves

(Continued on next page)

and ½ tsp. oil of cinnamon. Bring syrup to a boil. Drain off all liquid from rind and pour syrup over rind. Let stand in kettle overnight. Next morning drain off syrup, reheat and pour back over rind. Next morning repeat the same. Third morning, heat both rind and syrup. Pack in jars and seal. Makes about 8 pt.

Mrs. Lloyd Vance
Lincoln, Nebraska

Old Fashioned Watermelon Pickles

Watermelon rind	Sugar
Alum	Cinnamon
Salt	Cloves
Vinegar	Cardamom seed

Use thick rind melon (in fall striped ones best usually). Cut in 4-in. strips after peeling hard outer coat. Let stand overnight in water covering them and 1 tbsp. alum added and 1 tbsp. salt. Use dishpan, in morning, drain, cook in clear water slowly, until clear on edges. They will shrink as cooked. Drain off water. Make syrup of mild vinegar and sugar to cover rinds. Add stick cinnamon and a few cloves and cardamom seed. Simmer until clear rinds. Let stand overnight in liquid. Simmer again next day. Seal with liquid to cover. Have sugar enough so liquid is rather thick when dropping pickles in to cook.

Mrs. J. L. Scroggin
Oak, Nebraska

Watermelon Pickles

Watermelon rind	Sugar
Powdered alum	Oil of cinnamon
Vinegar	Oil of cloves

Rind from one watermelon peeled and cut into pieces. Put in large kettle, cover with water and boil until rind is clear (not too long--15-20 min.). Add 3 tbsp. powdered alum, stir well and let stand overnight. In morning, rinse with running water and drain. Make syrup of 1 qt. vinegar, 8 cups running water and drain. Make syrup of 1 qt. vinegar, 8 cups sugar, few drops oil of cinnamon and few drops oil of cloves (be careful not too much). Bring this to a boil, stir until sugar is dissolved, pour over pickles in a crock. Let stand overnight. Next morning, pour off liquid and heat to boiling, having added 1½ cups sugar. Pour over pickles. Repeat this 5 more days, adding 1½ cups sugar each day. Last day, pack pickles in hot jars and pour hot syrup over them and seal. Red and green food coloring can be added for color.

Mrs. Eliza Stephens
Elm Creek, Nebraska

Watermelon Pickles

Watermelon rinds
Salt
Alum
Vinegar
Sugar
Whole cloves
Whole black peppers
Stick cinnamon
Maraschino cherries

One large dishpan full and 1 gal. rinds. Cut off pink part and peel. Cut in quite large pieces and soak overnight in 1 gal. water and 3 tbsp. salt. Alum (size of walnut). Next day drain and wash well. Cover with fresh water and boil 1½ to 2 hr. Make syrup of 2 qt. vinegar, 7 lb. sugar, 4 level tsp. whole cloves, 2 level tsp. whole black peppers and 1 box of stick cinnamon. Boil 30 min. Add 1 large bottle of maraschino cherries and juice. Boil all together 45 min. Let stand in kettle overnight. Bring to boil next morning and can. These are delicious. Look very tasty on table.

Mrs. Josephine Jorges
Elwood, Nebraska

Watermelon Pickles

7 lb. melon
¼ tsp. alum
3 lb. sugar
1 pt. vinegar
½ tsp. oil of cloves
½ tsp. oil of cinnamon

Peel melon and cut in even-size pieces. Cook in clear water until tender. May add ¼ tsp. alum for crispness. Make syrup of sugar, vinegar, oil of cloves and oil of cinnamon. Pour syrup over pickles and let stand 4 days. Reheat syrup each morning and pour over melon. Fifth day, heat all together and put in jars and seal.

Mrs. A. J. Hindmand
Lincoln, Nebraska

Watermelon Pickles

1 medium-size melon
1 cup salt
7 cup sugar
½ tsp. oil of cinnamon
½ tsp. oil of cloves (from drugstore)
1½ pt. white vinegar

Cut watermelon rind in chunks. Soak overnight in salt water. Wash in morning. Cook in clear water 10 min. Make syrup of sugar, oil of cinnamon, oil of cloves and vinegar. Cook in syrup (boil up). Let stand overnight in crock. Repeat 5 days. Fifth morning, heat to boiling point and can. Makes 3 or 4 pt.

Mrs. Frank J. Hobbs
Lincoln, Nebraska

Watermelon Pickles

Watermelon rinds
Alum
Vinegar
Sugar
Cloves
Cinnamon

(Continued on next page)

Dissolve 1 tbsp. alum in water to cover 1 gal. of peeled and cut watermelon rinds. (Cut into size of pickles wanted). Let stand overnight, drain, cover with cold water, bring to boil, cool in same water, drain. Boil 1 pt. vinegar, 4 pt. sugar (½ oz. cloves and 1 oz. cinnamon in cloth bag) and pour over pickles. Each morning for 4 days, drain off liquid, bring to boil and pour back on pickles. Last morning, bring pickles to boiling point and can. They will keep in stone jar. This is a very tasty and very firm pickle, the nicest we have ever eaten. We got the recipe from Mrs. J. C. Hagey at Osceola, Nebraska, 40 years ago and it is the recipe we always use for watermelon pickles.

Hilda Buettner
Madison, Nebraska

Watermelon Pickles

Watermelon rinds
5 cups sugar
2 cups white vinegar
½ tsp. oil of cinnamon
¼ tsp. oil of cloves

Pour boiling water over melon rinds and boil until tender. Drain well. Boil sugar, vinegar, cinnamon and cloves. Put well-drained rinds into syrup while hot. Reheat to boiling point for 2 days. Third day, heat, put in jars and seal.

Mrs. Charles Bors
Wilber, Nebraska

Watermelon Rind Pickles

Watermelon rind
1 pt. vinegar
7½ cups sugar
½ tsp. oil of cinnamon
½ tsp. oil of cloves
Salt

Cut off all green and red of watermelon rind. Then cook in slightly salted water until done--but not too soft. Drain off water and put into dish. Pour syrup over rind. Next morning, drain off syrup and bring syrup to boil. Pour over rinds. Do again 3rd morning. Fourth morning, bring all to boil or pour off syrup, bring to boil and pour over rinds which have been packed into jars and seal.

Mrs. E. F. Koch, Jr.
Omaha, Nebraska

Spicy Watermelon Pickles

2 lb. watermelon rind
¼ cup pickling salt
5 cups sugar
2 cups white vinegar
1 lemon (sliced thin)
½ tsp. oil of cinnamon
¼ tsp. oil of cloves
Ginger root (if available)

Pare rind and cut into 1-in. cubes. (Dark green melons make best pickles). Soak rind overnight in salt water (¼ cup pickling salt to 1 qt. water). In morning, drain it well, then rinse and cover with fresh cold water. Cook until just tender. Drain well. Make syrup of sugar, vinegar, lemon, oil of cinnamon, oil of cloves and 2 cups

(Continued on next page)

water. Simmer syrup for 10 min. to blend, then drop in rind. Simmer until rind is clear. Cover and let stand overnight. Do this 2 or 3 days, then can hot in sterile jars, filling to ½ in. from top. Seal. (If I can get ginger root, I like to toss 4 or 5 small pieces in the syrup with the other spice oils).

Mrs. Fred Arnholt
Thedford, Nebraska

Sweet Pickled Watermelon Rind

7 lb. watermelon rind
Alum
1 qt. cider vinegar
3 tbsp. stick cinnamon (broken)
½ tbsp. whole cloves
2 tbsp. white mustard seed

Use about 7 lb. rind. Cut white portion of rind into uniform pieces. Cover with alum solution (2 tsp. alum to each qt. of water). Let stand overnight. In morning, drain, cover with slightly salted water, bring to boil and cook 10 min. Drain, chill thoroughly, let stand in ice water 2 hr. Drain and dry on cloths. Make syrup of sugar, vinegar, cinnamon, cloves and mustard seed (Tie spices loosely in cloth). Simmer rind in syrup until tender. Pack in sterilized jars, covering with syrup. Mature zucchini squash may be used instead of watermelon rind. Discard seed portion of squash.

Mrs. H. O. Werner
Lincoln, Nebraska

MISCELLANEOUS

Always Ready or Lazy Wife Pickles

Cucumbers
Salt
Sugar
Cider vinegar
Cloves
Cinnamon
Allspice

Gather cucumbers every day. Wash carefully. Pour boiling salt water (1 cup salt to 1 gal. water) over them. Let stand 1 hr. Take out of salt water, wipe dry, drop into jars. On each qt. of pickles put 2 tbsp. sugar. Cover with strong cider vinegar to which cloves, stick cinnamon and allspice have been added. Seal tightly and they will be ready to use in 10 days or 2 wk.

Earl May Seed & Nursery
Shenandoah, Iowa

Apple Pickles

Whitney crabs (or red jelly crabs)
2 cups vinegar
4 cups sugar
2 sticks crushed stick cinnamon
1 tsp. whole cloves

Drop peeled or unpeeled Whitney crabs (or red jelly crabs) into boiling liquid and cook until clear. Seal with boiling hot syrup covering apples in sterilized jars. Make liquid of 2 cups vinegar, 4 cups sugar. Season with 2 sticks crushed stick cinnamon and 1 tsp. whole cloves tied in a cloth (or a few drops of oil of cinnamon and oil of cloves).

Mrs. Carol G. Lutz
Arcadia, Nebraska

Bean Pickles

Beans
Salt
Sugar
Mustard seed

Cook full-length beans in salted water until done. Pack carefully lengthwise in jars. Cover with boiling liquid in proportions of 1 cup vinegar, 1 cup water and 1 cup sugar. Add a sprinkle of mustard seed. Seal.

Mrs. Carol G. Lutz
Arcadia, Nebraska

Beet Marbles

Beets
2 cups sugar
2 cups vinegar
½ cup water
¼ tsp. non-iodized salt

Bring small beets to boiling, reduce heat and simmer (covered) until tender 40-45 min. Skin. Cover with syrup made of sugar, vinegar, water and salt. Bring to boil. Add beets, reduce heat and let remain over low heat for 20 min. but do not boil. Then put beets in hot sterile jars, fill with the hot syrup to ½ in. of top and seal.

Mrs. Fred Arnholt
Thedford, Nebraska

Beet Pickles

Beets
2 cups sugar
2 cups vinegar
4 cups water

Cook beets, slip skins, cut in desired shape, put in qt. jars. Pour over hot solution of sugar, vinegar and water. Makes about 5 qt. Seal. Pressure 10 min. at 5 lb.

Wilma Nyffeler
Alliance, Nebraska

Beet Pickles

Beets
1 cup sugar
1 cup vinegar
1 cup "strained" water

Bring cooked peeled beets to a simmer in liquid in the proprotions of 1 cup vinegar and 1 cup strained water the beets were boiled in. Seal in sterilized jars.

Mrs. Carol G. Lutz
Arcadia, Nebraska

Beet Pickles

Beets
2 cups vinegar
2 cups sugar
2 cups water
1 lemon sliced
1 tsp. cinnamon
1 tsp. cloves
1 tsp. allspice

Mix vinegar, sugar, water, lemon, cinnamon, cloves and allspice and bring to boil. (Spices should be in bag). Have beets cooked and skinned. Put in syrup. Bring to boil and can.

Mrs. Eliza Stephens
Elm Creek, Nebraska

Beet Pickles

Beets
3 cups sugar
3 cups vinegar
1 cup water
Pickling spices

Cook beets until tender. While hot, peel and slice and put in can while hot. Boil sugar, vinegar, water and pickling spices and pour over beets while hot.

Mildred Saf
Oakland, Nebraska

Beet Pickles

Beets
2 cups sugar
1 cup cider vinegar

Cook beets until tender. Slice and put in jars. Boil sugar and vinegar. Pour over beets and seal.

Mrs. C. A. Valder
Tekamah, Nebraska

Beet Pickles

Beets
1 qt. vinegar
1 qt. water
1 qt. sugar
3 tsp. cinnamon
1 tsp. salt

Wash, cook, peel and slice beets. Boil together the vinegar, water, sugar, cinnamon and salt. Add beets, again bring to boil and seal in jars.

Mrs. Clair Blezek
Randolph, Iowa

Beet Pickles

Beets
1 qt. vinegar
1 qt. sugar
1 qt. water (in which beets have been cooked)
3 tsp. powdered cinnamon
3 tsp. salt

Wash and top beets. Cook until tender. Peel or slip skins. Pack in jars while hot. Cover with boiling solution of vinegar, sugar, water, cinnamon and salt. Boil all together. Cover beets, seal and set away. May be used in 2 or 3 wk.

Earl May Seed & Nursery
Shenandoah, Iowa

Beet Pickles

Beets
Salt
2 cups vinegar
3 cups sugar
1 cup water

Boil beets in salted water until tender. Remove skins and slice. Bring vinegar, sugar and water to boil. Pour over beets and bring to boil and can. No spices. About 2 doz. medium-size beets should make 3 qt. of pickles.

Mrs. Carl B. Hoy
Vermillion, So. Dakota

Beet Pickles (Sour)

Beets
2 cups vinegar
1 cup sugar

Cook beets, slip skins, cut in quarters, put in sterile hot pt. jars. Cover with solution (hot) of vinegar and sugar. Seal. Do not add water.

Wilma Nyffeler
Alliance, Nebraska

Canned Peppers

Peppers
1 pt. vinegar
1 pt. water
4 tbsp. salt
4 tbsp. sugar

Cut peppers into strips. Pour boiling water over them. Let stand few minutes. Drain, pack into jars. Boil vinegar, water, salt and sugar. Pour over peppers and seal.

Mrs. Don Keetle
Ashland, Nebraska

Carrot Pickles

2 to 3 bun. small carrots
1 cup sugar
2 cups vinegar
1½ cups water
1 tsp. salt
1 stick cinnamon
1 tbsp. mixed pickling spices

Cook carrots until just tender. Drain and remove skins. Leave small carrots whole. Cut larger ones into desired pieces. Combine sugar, vinegar, water and salt. Tie spices in cheesecloth bag. Add to vinegar mixture. Boil 5 to 8 min. Pack carrots into hot jars, leaving ½ in. head space. Remove spice bag. Heat syrup to boiling. Pour hot over carrots, leaving ½ in. head space. Adjust caps, process 30 min. in boiling water bath. This recipe makes 3 pt.

Mrs. Stanley Cada
Schuyler, Nebraska

Cherry Olives

1 qt. cherries with pits and stems
1 pt. vinegar
1 tsp. salt
1 pt. water

Prepare brine, add vinegar. Put cherries in jar, pour liquid (made of vinegar, salt and water) on cherries and seal. Let stand 1 mon. before using.

Paul Bluhm
Seward, Nebraska

Crab-apple Pickles

2 cups vinegar
4 cups sugar
2 cups water
Spices

Wash and cut out the blossom end of apples. Be sure to leave the stem on. Prick each apple with fork several times. Cook the apples in clear water until nearly done. Then drain. Boil vinegar, sugar, water and spices to taste. The water the apples were cooked in can be used. Put the cooked apples in the syrup and bring to a boil. Seal.

Mrs. William J. Murphy
O'Neill, Nebraska

Cucumber Pickles

Cucumbers
Salt
Vinegar
2 minced onions
20 cloves
1 oz. mustard seed
1 oz. celery seed
Few blades of mace
1 cup sugar

This recipe is one which we cut out of our old July 1, 1913, Nebraska Farm Journal. Choose only small cucumbers, as they make nice looking as well as tender pickles. Lay 150 small cucumbers in cold water for an hr., then drain and turn into a stone crock. Pour over them a cold brine strong enough to float an egg. Let stand 3 days, then drain and dry on a towel, then let soak until the next day in cold water. Have ready on the stove hot vinegar in which you have boiled onions, cloves, mustard seed, celery seed and mace. Fill the jar with this boiling mixture and add sugar, stirring the cucumbers up from the bottom. Cover tightly. In a week scald the vinegar again and return to the jar. Let the pickles stand for 6 wk. before using; 6 mo. is even better. Fannie H. Kamfren, Edmunds County, So. Dakota.

Mr. and Mrs. George Cherny
Beatrice, Nebraska

Dilled Beans

Full-length beans
1 head of dill
1 small clove of garlic
2 cups vinegar
½ cup salt
2 cups water

Boil beans 5 min. Cook--pack in jars with dill and garlic. Heat to boiling the vinegar, salt and water. Pour over packed beans in jars and seal.

Mrs. Carol G. Lutz
Arcadia, Nebraska

Dilled Zucchini

6 lb. zucchini (about 16 cups)
2 cups celery (sliced thin)
2 large onions (2 cups)
1/3 cup salt
Ice cubes
2 cups sugar
Fresh dill (or 2 tbsp. dill seed)
2 cups white vinegar
Garlic

Makes 12 8-oz. jars or 6 16-oz. jars. Mix zucchini (trimmed and sliced thin), celery, onions (chopped) and salt in large bowl. Place layer of ice cubes on top. Cover. Let stand 3 hr. Drain well. Combine sugar, dill and vinegar in kettle. Heat, stirring constantly, to boiling. Stir in vegetables. Heat, stirring several times, just to a full rolling boil. Ladle into hot sterilized jars. Place 1 or 2 pieces of garlic in each. Seal, following manufacturer's directions. Cool jars, label and date. The pickles taste a bit like a cucumber but what a delightful change!

Mrs. H. J. Newman
Alliance, Nebraska

Dilly Beans

Green beans
Ground red pepper
Garlic cloves
Dill heads
Cider vinegar
Salt (not iodized)
Water

Pack washed, steamed (3 min.) green beans. To each pint add ½ tsp. ground red pepper, 1 clove garlic, 1 dill head. Heat 2 cups pure cider vinegar, 2 cups water ½ cup pure salt (not iodized). Boil, pour over steamed beans and seal. 4 lb. beans make 6 pt.

Mrs. Frank J. Hobbs
Lincoln, Nebraska

Dilly Beans

Green beans
Ground red pepper
Garlic cloves
Dill heads
Cider vinegar
Salt (not iodized)
Water

Pack in jars, washed and steamed greenbeans (steam beans 3 min.). To each pint add ½ tsp. ground red pepper, 1 clove garlic, 1 dill head. Heat 2 cups pure cider vinegar, 2 cups water, ¼ cup pure salt (not iodized). Boil. Pour over beans in jars and seal. 4 lb. beans (young and tender) make 6 pt. These are very good with a Dutch brunch.

Mrs. Frank J. Hobbs
Lincoln, Nebraska

Freeze Fresh Cucumbers

Cucumbers
½ tsp dill seed
1 tsp. sugar
1/8 tsp. salt
½ cup white or wine vinegar

Peel and slice cucumbers and freeze dry. Cover with solution (when you are ready to use them) made of dill seed, sugar, salt and vinegar. Pour over while these are frozen and let them thaw in it.

Mrs. I. G. Saf
Oakland, Nebraska

Frozen Cucumber Pickles

2 gal. fresh medium-size cucumbers
1 cup coarse salt
1 tsp. powdered alum
4 qt. white sugar
2½ qt. white vinegar
5 or 6 sticks cinnamon
Grape leaves

(Continued on next page)

Slice cucumbers ¼-in. thick. Add salt and cover with boiling water. Let stand 7 days. Drain and wash on 8th day. Cover again with boiling water. Ninth day, drain, add alum and cover with boiling water. Tenth day, drain, boil together the sugar, vinegar and cinnamon. Pour over pickles. Let stand overnight, drain, boil the syrup and pour over pickles 3 more days. Pour the boiling syrup over the pickles packed in jars and seal. Grape leaves may be laid over top of pickles the 10th day to help retain color.

Mrs. M. M. Wright
Benedict, Nebraska

Harvest Pickle

Dill-size cucumbers
Salt
1 cup vinegar
2 cups sugar
¼ tsp. turmeric
1 tsp. mixed pickle spices
1 tsp. celery seed

Pour boiling water over cucumbers each morning for 6 mornings. Put fresh boiling water on each day. Seventh day, put enough salt in boiling water to taste. Eighth day, cut pickles in chunks and put in jars and cover with syrup made of vinegar, sugar, turmeric, pickle spices and celery seed. Put spices in small bag and boil with rest. Seal while hot.

Mrs. Henry Paeper
Nickerson, Nebraska

Hunters Pickles (unsealed)

1 doz. hot dogs
Pickles cut into halves
Hot pickle juice from commercial canned hot peppers

Pack in jars and let set for 1, 2, 3, etc. months. When needed for hunting or fishing put in jars and screw the lids on tight. They will keep several days. We use them without or with the liquid poured over them before sealing with a lid and they do not require ice. If the pepper juice is too hot, I usually add vinegar and water in equal portions for curing. We also use the hot dogs alone in the juice or the pickles alone. We use venison bologna, beef bologna or summer sausage (cured). Got that everlasting sandwich beat a mile for warm weather.

Mrs. Paul Harrison
Newport, Nebraska

Ice Pickles

Pickles
Onion slices
Celery seed
1 cup sugar
½ cup salt
1 qt. vinegar

Soak pickles in ice water for 2 hr. Dry with a cloth and pack in jars. (If you like, put in a few onion slices and celery seed). Let sugar, salt and vinegar come to boil and pour over pickles in jar. These are very crisp and brittle, due to action of boiling vinegar over icy pickles.

Mrs. George Schmid
Monroe, Nebraska

Ice Water Pickles

6 lb. medium-size cucumbers
Ice water
Pickling onions
Celery
Mustard seed
3 cups sugar
3 qt. white vinegar
1 cup salt

Ice water pickles are crisp and delicious. This recipe is an old favorite. Cut each cucumber in 4 to 8 pieces, according to size of cucumber. Soak in ice water 4 to 5 hr. Drain. Pack into sterilized qt. jars. Add 6 pickling onions, 1 piece celery and 1 tsp. mustard seed to each jar. Prepare solution of sugar, vinegar and salt. Bring to boil, pour over cucumbers and seal jars. This makes 6 qt. Ready in 2 to 4 wk.

Mrs. William R. Wessel
Roca, Nebraska

Icicle Pickles

1½ gal. cucumbers
1 pt. coarse salt
Alum lump size of walnut
2½ pt. vinegar
12 cups sugar
6 to 8 sticks cinnamon

Wash cucumbers and cut in strips. Use cucumbers with small seeds. Then put brine over them. Make brine of 1 pt. coarse salt to 1 gal. water. Pour over pickles, let stand overnight, drain. Make alum water of lump of alum size of walnut and enough boiling water to cover. Let stand 1 day. Take out of alum water and pour over cucumbers a syrup made of vinegar, sugar and cinnamon sticks. Heat syrup to boiling and pour over pickles each day for 4 days. Just heat the syrup. Can on 4th day with boling syrup. These pickles are very crisp.

Mrs. Lola Layman
Fremont, Nebraska

Icicle Pickles

Cucumbers
Onion slices
1 qt. white vinegar
2½ cups white sugar
1 tbsp. salt (not iodized)
1 tbsp. celery seed
Powdered alum

Slice cucumbers lengthwise or leave whole. Cover with cold water overnight. Drain. Place 2 slices onion in bottom of each jar. Fill jars with (cold) cucumbers and place a slice of onion on top. Pour syrup (made of vinegar, sugar, salt and celery seed) over cukes boiling hot. Dilute vinegar or add sugar according to taste. Add powdered alum about size of pea to each jar. Seal.

J. O. Olson
Lincoln, Nebraska

Italian Cucumber Pickle

5 doz. green cucumbers (4 in. long)
5 cups vinegar
1½ cups salad oil
½ tsp. mustard seed (in each jar)
20 small onion (sliced)
2 tbsp. celery salt
Salt

Wash cucumbers, slice, sprinkle freely with salt. Let stand 24 hr. Add onions. Combine other ingredients and beat well. Pour over onions and cucumbers which have been packed in sterilized jars. Seal at once. Ready in 6 wk.

The Noll Sisters
Lincoln, Nebraska

"Jugged Rabbit"

Rabbit
Lard
Sweet herbs
Onion
Cloves
Shallots
Pepper
Salt
Paste

Cut a rabbit into pieces, rub with lard, put in a jug or earthen jar, with the blood of the rabbit, a bundle of sweet herbs an onion stuck with cloves, a few shallots, pepper and salt. Cover mouth of jar with common paste, put into pan of boiling water. Boil slowly 4 hr.

Mrs. Floyd Dreyer
Cook, Nebraska

Lazy Housewife Pickles

Cucumbers
3 cups sugar
2 cups water
½ cup coarse salt
1 qt. vinegar
Onion slices
2½ tsp. celery seed
2½ tsp. dill seed (or dill heads)
Alum
5 garlic buds

Slice cucumbers lengthwise, enough for 5 qt. Place 2 pieces (or 1 piece) of onion on top of each jar. Add celery seed, dill seed and alum (½ tsp.) to each jar or 1 dill head, 1 garlic bud and pea-size alum to each jar. Cucumbers may be sliced, chunked or split.

Mrs. Floyd Dreyer
Cook, Nebraska

Lazy Housewife Sweet Dill Pickles (5 qt.)

4 cups sugar
½ cup salt
1 qt. vinegar
1 pt. water
20 slices onion
2½ tsp. celery seed
2½ tsp. dill seed
Fresh cucumbers
5 pieces of alum half as large as a pea

Fill 5 jars with fresh cucumbers. (I use powdered alum about 1/8 tsp. per jar). Place 4 slices of onion in bottom of each qt. jar. Slice cucumbers lengthwise and pack in jars. Heat vinegar water, salt and sugar mixture to boiling. Fill jars. To each jar add ½ tsp. dill seed or some dill, ½ tsp. celery seed and the alum. Seal. Pickles will be ready to eat in about 3 wk.

Mrs. Arthur H. Johnson
Red Oak, Iowa

Lazy Wife Pickles

1 peck medium-size cucumbers
1 gal. vinegar
1 cup salt
1 cup dry mustard

These pickles are so easily made as to be amusing and they are exceptionally delicious. To vinegar add salt and mustard. Put this in a stone crock. Wash cucumbers. Dry thoroughly (this is very important). Drop cucumbers into the vinegar and put lid on tight. In morning, pickles are ready to be eaten. You may add cucumber from time to time to this same brine so that you may have these delicious pickles all summer. Be sure your cucumbers are dry when dropped into the brine.

Mrs. William R. Wessel
Roca, Nebraska

Lazy Woman Pickles

Cucumbers
Pure pickling salt
White vinegar (5 or 6% acidity)
Dill seed (or dried dill tops)
Sugar
Garlic
Juice from hot pickled peppers

Scrub enough cucumbers to fill your jars. Allow half as much liquid as your jar has size (1 qt. needs 1 pt.). Use 1 tbsp. salt per qt. and be sure the salt is pure salt. Use pickling salt only. Liquid is white vinegar (5 or 6% acidity). Dissolve salt in vinegar and pour on cucumbers. Seal. Late in season or winter when you have time these may be opened and made into any style pickle desired. They are crisp, cured and keep without spoiling. When made into sweet pickles they stay crisp. To make into dills, use half water and half vinegar with dill seed (or dried dill tops). You can also add sugar, garlic, juice from hot pickled peppers or any other seasoning that is liked for dills. I have used the vinegar boiling hot over the pickling cucumbers in the jars with the salt but I add 1 tsp. sugar. I usually use the vinegar cold, however. Salt--pure pickling salt. Vinegar--5 or 6% acidity and always white vinegar.

Mrs. Paul Harrison
Newport, Nebraska

Mixed Pickles

3 qt. sliced cucumbers, unpeeled
1 qt. sliced onions
3 large green peppers, cut in strips
1 cup salt
3 cups sugar
3 cups vinegar
1 tbsp. mustard seed
1 tbsp. celery seed
1 tsp. turmeric powder
1 small can chopped pimento

Cover cucumbers, onions and peppers with salt water and let stand overnight. Drain. Make syrup of rest of ingredients and bring to boil. Add cucumbers, onions and peppers. Bring to boil again and cook 2 min. Seal hot. Makes 5 pt.

Mrs. Clarence Meyer
Lincoln, Nebraska

Mixed Pickles

1 small cauliflower head
Small onions
Peppers (cut in strips)
Cucumbers (cut in cubes)
Bean (optional)
Cider vinegar
Salt
Sugar
Mustard
Mixed spices

Scald cauliflower, onions, pepers, cucumbers and beans (optional) in salt water until tender. Then drain until dry and pack into wide-mouth jars. Boil in each pint of cider vinegar--1 tbsp. sugar, ½ tsp. salt, 2 tbsp. mustard and (tied in bag) 2 tbsp. mixed spices. Pour entire mixture of ingredients over pickles and seal. A sweet vinegar such as used for sweet pickles may be used in the same manner.

Earl May Seed & Nursery Co.
Shenandoah, Iowa

Olive Oil Pickles

1 gal. medium and large cucumbers
1 cup coarse salt
1 qt. sliced onion
1 cup vinegar
1 cup sugar
1 tbsp. mustard seed
1 tbsp. celery seed
½ cup olive oil

Slice cucumbers rather thick. Stir in salt and let stand overnight. Drain and wash off salt. Put in a kettle and add sliced onion, vinegar, sugar, water, mustard seed and olive oil. Let boil 20 min. Put in sterilized jars and seal tightly. Makes 7 pt. Ready to use as soon as cool.

Earl May Seed & Nursery
Shenandoah, Iowa

Better Homes and Gardens Olive Oil Pickles

5 qt. sliced cucumbers (about 25 4-5" long)
½ cup sliced onions
½ cup coarse salt
2½ tsp. powdered alum
2 cups vinegar
2 cups water
¼ cup white mustard seed
2 tbsp. black mustard seed
1½ tsp. celery seed
3/4 cup olive oil

Slice cucumbers thin and arrange in layers with sliced onion. Sprinkle each layer with salt. Let stand overnight. Drain, dissolve alum in half the vinegar and water solution, pour over pickles. Let stand 4 hr. Drain and combine pickles with seasonings. Pack in hot sterilized jars. Mix remaining vinegar and water with oil and beat until well-blended. Don't heat. Pour over pickles and seal. Makes about 8 pt.

Earl May Seed & Nursery
Shenandoah, Iowa

One Quart of Pickles

Pickles
Onion
Green pepper
Sugar
Vinegar
Pickling salt

Slice washed pickles in a jar until half full. Slice one onion in it, one green pepper diced or in rings. Mix 1½ cups sugar, 1 cup vinegar, 1 tbsp. pickling salt. Bring to boil, pour on and seal.

Mrs. C. A. Valder
Tekamah, Nebraska

One Quart Pickles

Cucumbers
Onion slices (or real small onions)
Green pepper slices
1½ cup sugar
1 cup vinegar
1 tbsp. pickling salt
(Pickling spices if desired)

Mix sugar, vinegar and pickling salt (spices if desired) and bring to rolling boil. Have washed and sliced cukes in jars, not too packed. Few slices of onion or real small onions, also green pepper slices if desired. Pour hot syrup over cukes and seal jars. I used pint jars. Canbe used at end of 4 days.

Mrs. Emil Walla, Sr.
Linwood, Nebraska

Pickled Beans

Beans
1 cup sugar
2 cups vinegar
2 cups water
1 tbsp. salt
1 tbsp. celery seed

Cook beans in salt water until tender. Drain, about 4 qt. Boil sugar, vinegar, water, salt and celery seed. Boil syrup 5 min., add cooked beans and cook 5 more min. Pack in jars and seal.

Mrs. Frank Nickman
Pleasanton, Nebraska

Pickled Beans

Beans (yellow preferred)
Salt
2 cups water
1 cup vinegar
1 cup sugar
Onion
Green dill heads

Boil clean (yellow preferred) beans in salt water until half done. Drain. Bring to boil water, vinegar, sugar and onion to taste. Cook beans 15 min. in vinegar syrup. Place a head of green dill on bottom of each jar. Pack beans and cover with vinegar syrup. Seal while hot.

Mrs. Stanley Cada
Schuyler, Nebraska

Pickled Beets

Beets
1 cup vinegar
1 cup water beets were cooked in
Stick cinnamon

Cook desired amount of cleaned beets until tender. For syrup combine vinegar, sugar and "beet water." Bring to boil and add beets. Boil 5 min. Pack into jars and process 5 min. in hot water bath, then complete, seal. Stick cinnamon may be added to jar if desired.

Mrs. Roy Davis
Curtis, Nebraska

Pickled Beets

2 cups sugar
1 tbsp. whole allspice
1½ tsp. salt
2 stick cinnamon
1 tsp. cloves
3½ cups vinegar
1½ cups water
3 qt. peeled, cooked, small beets (or may slice or quarter them).

When cooking beets, wash well, leave about 2 in. of stems on beets. Cover with boiling water and cook till tender. Combine all ingredients except beets and simmer 15 min. Pack beets in sterilized jars. Leave ½ in. head space. (Either strain off spices or cook in cloth bag). Bring liquid to boil. Pour boiling hot over beets, leaving ½ in head space. Adjust caps. Process pints and quarts 30 min. in boiling water bath. Yields about 6 pt. These are very good. I like to use the jar caps that are white enamel for pickles (Ball Brand).

Mrs. Frank J. Hobbs
Lincoln, Nebraska

Pickled Beets

Medium-size beets
Salt
Vinegar
Sugar
Water
Stick cinnamon pieces
Cloves

Wash medium-size beets and cook with roots left on and leaves cut to leave 2 in. on top of beets. Cook uncovered until almost tender. Cold dip, remove skins, tops and roots. Add 1 tsp. salt to each qt. and vinegar (not too strong) or make pickling syrup of 2 cups sugar, 2 cups vinegar, 2 cups water, 1 tbsp. stick cinnamon pieces and 1 tsp. cloves (tie cinnamon and cloves in cheesecloth bag). Boil solution 10 min. and pour boiling hot over beets. Seal. Golden wax beans, stems of swiss chard or small carrots may be pickled in same way.

Mrs. S. C. Huffman
Lincoln, Nebraska

Pickled Beets

Beets
2 cups vinegar
1½ cups sugar
1 cup water beets were cooked in

Wash beets and peel them. Cut them up ready to can. Cook until done. Let vinegar, water and sugar come to boil. Add hot cooked beets and let come to boil again. Can hot. Makes 2 qt.

Mrs. Frank Hahn
Johnson, Nebraska

Pickled Green Beans

Green beans
Dill

Green beans are very good dilled. They should be small-size beans and cooked slightly before pickling.

Esther Bruneau
Jefferson, So. Dakota

Pickled Onions

1 gal. onions (pearl, small-size)
1 cup salt
2 cups sugar
1½ qt. white vinegar
3 tbsp. white mustard seed
3 tbsp. whole pepper
Salt

Cover onions with boiling water. Let stand 2 min. Dip in cold water and peel. Sprinkle with salt. Cover with cold water. Let stand overnight. Rinse, drain, put in jars. Simmer spice 15 min. Cook 3 min. Pour hot over onions and seal.

Mrs. William R. Wessel
Roca, Nebraska

Pickled Onions

Small, white onions
Mixed spices
Sugar
Vinegar

Blanch small, white onions in boiling water 2 min. Dip in cold water and remove skins. Pack in qt. jars. For each qt., tie 1 tbsp. mixed spices in cloth bag. Dissolve 1¼ cup sugar in 1 cup vinegar and 1 cup water. Add spices and simmer 40 to 60 min. Pour hot solution over onions and seal.

Mrs. Stanley Cada
Schuyler, Nebraska

Pickled Walnuts

Walnuts
Salt
Vinegar
Sugar
Whole cloves
Allspice
Peppercorns
Mace

Gather black walnuts when soft enough to be pierced by a needle in July. Prick each with a large needle well through, holding in a cloth to avoid staining hands. Cover with strong salt water (1½ pt. salt to 1 gal. water). Let stand 2 or 3 days, changing brine every day. Then pour over them a brine made by dissolving salt in boiling water (let it get cold before using). Let stand 3 days, renew brine and let it stand 3 days more. Now drain and expose to the sun for 2 or 3 days or until they become black, or put in cold water for half a day, and pack in jars not quite full. Proportions are a 100 walnuts to each gal. of vinegar. Boil vinegar 8 min. with a teacup sugar, 3 doz. each whole cloves and allspice, 1½ doz. peppercrons and 1 doz. blades of mace. Pour vinegar over walnuts scalding hot. In 3 days draw off vinegar, boil and pour over walnuts again while hot and at end of 3 days repeat process. They will be fit to eat in a month and will keep for years.

Mrs. Niels Miller
Herman, Nebraska

Pickles or Beets

Cucumbers (or beets)
2 cups water
1 cup sugar
1 cup vinegar
Dill (or spice)

Let cukes stand in salty water overnight. Cook pickles until they lose their color only. Put into jars and seal.

Bessie Meduna
Weston, Nebraska

Pickling Syrup

4 cups sugar
2 cups water
1 lemon sliced thin
2 cups vinegar
1 tsp. cloves
1 tsp. allspice
1 tbsp. cinnamon

Mix and boil 10 min. While boiling hot pour over pickles. This recipe can be used for cucumber, pears, crabapples and peaches.

Mildred Saf
Oakland, Nebraska

Pimiento Pickles

6 qt. sliced cucumbers
12 medium-size onions (sliced)
1 can pimientos (or 5 fresh pimientos)
2 green peppers
1 small cup salt
1½ qt. vinegar
¼ cup white mustard seed
3 tsp. powdered turmeric
6 cups sugar

Mix cucumbers, onions, pimientos, peppers and salt. Let stand 3 hr. Drain well. Mix and bring to boil vinegar, mustard seed, turmeric and sugar. Add pickle mix and heat. Seal.

Mrs. William J. Murphy
O'Neill, Nebraska

Pink Cucumber Pickles

Ripe yellow cucumbers
Salt
Alum
Sugar
Vinegar
Whole cloves
Cinnamon
Maraschino cherries
Red food coloring

Remove rind from cucumbers and cut cucumbers in 1 in. or so pieces. Soak several hr. in salt water. Drain off brine and cook until tender in fresh water to which ½ tsp. alum is added during last 10 min. of cooking. For every 5 lb. of pieces, make syrup of 3 lb. sugar, 1 pt. vinegar, 2 tsp. whole cloves and ½ tsp. cinnamon. Boil 15 min. Add pickles and simmer 10 min. Add 1 small bottle maraschino cherries. (Add a few drops of red food coloringif not pink enough). Simmer few minutes. Seal in jars.

Mrs. Helen M. Kucera
Hastings, Nebraska

Pink Pickled Turnips

Small turnips (size of pigeon eggs)
3/4 cup salt (pure, not free-running nor iodized)
1½ qt.vinegar (5% acidity and white)
Small slices garlic (1 slice per jar)
Small whole beets cut in halves (cooked or raw; 1 piece per jar)
2½ qt. water (soft only)

Scrub turnips and beets. Pare tops and bottoms only. Beets may be left raw as they are for color only. Place a piece of beet in bottom of each jar and fill jar with whole raw turnips to the half-way mark. Add dill and garlic--finish filling the jar. (These may be put in crock--covered with grape leaves and at the end of a couple of months they will be ready. They must be weighted with a plate over the top and a sack (of plastic) filled with the pickle juice to hold the plate down. Either way makes good pickles. Boil the salt, water and vinegar 10 min. Cool. Pour over turnips in jars or crock, seal jars or weight turnips in crock. Cover and set aside for a month. (Jars will take about 2 months). Discard beets when using pickles. This was Mrs. Ferris Abdella's recipe that she brought from Syria when she came to Stuart as a young married woman.

Mrs. Paul Harrison
Stuart, Nebraska

Red Beet Pickles

Beets
2 cups sugar
2 cups water
2 cups vinegar
1 tbsp. cinnamon
1 tsp. cloves
1 tsp. allspice

Cook beets until tender. Cold-dip and slip skins. Make pickling syrup of sugar, water, vinegar, cinnamon, cloves and allspice. Cover beets with mixture and simmer 15 min. Seal. (Use the water the beets were cooked in but run through a milk filter). This gives beets a redder color. This recipe can be used to pickle beans, carrots and Swiss chard stems.

Mrs. William J. Murphy
O'Neill, Nebraska

Sauerkraut Salads

No. 1

1 large can sauerkraut (drained well)
1 cup celery (chopped fine)
1 small can pimiento
½ tsp. celery seed
½ tsp. mustard seed

Optional ½ cup salad oil, in ½ cup vinegar, 3/4 cup sugar, mix well, pour on above and mix all together well. Let stand in refrigerator all night.

No. 2

1 large can sauerkraut (drained well)
2 cups diced celery
1 cup diced onion
1 large green pepper (diced, seeded)
3/4 cup sugar

Chill overnight.

No. 3

1 large can sauerkraut (drained well)
1 green pepper (chopped fine)
1 medium onion (chopped fine)
3 stalks celery (chopped fine)
1 cup chili sauce
3/4 cup brown sugar
1 tsp. paprika
3 tbsp. lemon juice

Drain kraut, put in pepper, celery, onion, mix well. Mix other ingredients together, then mix all together. Let stand overnight.

Mrs. C. A. Valder
Tekamah, Nebraska

Sour Pickles--Cold

Cucumbers
2 tbsp. sugar
2 tbsp. salt
2 tbsp. dry mustard
1 pt. vinegar

Put small cucumbers in sterilized jar. Pour mixture over and seal.

Mrs. G. C. Albaugh
Lincoln, Nebraska

Spiced Crabapples

4 lb. crabapples
2½ cups white vinegar
2 cups sugar
4 cups water
4 cups granualted sugar
1 tbsp. whole cloves
3 3-in. cinnamon sticks
1 tsb. whole ginger

Wash and remove blossom ends of crabapples. Prick apples several times to retard bursting of skins. Heat vinegar, water and sugar to boiling. Add spices tied in cheesecloth bag. Cook part of crabapples in syrup 2 min. Remove and add more crabapples and cook, repeating until all crabapples are used. Pour syrup over carbapples with spice bag and let stand overnight. Remove spice bag. Pack apples into hot, clean jars. Heat syrup to boiling. Pour over fruit. Adjust covers according to type of lid used. Process 30 min. in boiling water bath. Makes 4 to 5 pt.

Mrs. Val Kuska
Lincoln, Nebraska

Spiced Peaches

6 lb. peaches (about 36 small ones)
Cloves (1 for each peach)
Cinnamon oil (1 dram for each jar)
3 cups white Karo syrup
3 cups sugar
2 cups white vinegar
1 piece ginger root

Put 3 cups white Karo syrup, 3 cups sugar, 2 cups white vinegar, 1 piece ginger root in saucepan and bring to boil. Peel 6 lb. of peaches (about 36 small ones). Drop into boiling syrup a few at a time. Cook until tender. Let peaches stand in syrup overnight. Stick 1 clove into stem end of each peach. Pack into hot, sterile jars, and add 1 drop oil of cinnamon to syrup for each jar of pickles you have. Boil down syrup until you have 3-4 cups left. Fill jars to within ½ in. of top, seal immediately and process 15 min. in boiling water bath. Yield: 5-6 pt.

Mrs. Fred Arnholt
Thedford, Nebraska

Spiced Peaches

4 lb. peaches (16 medium)
4 cups granulated sugar
1½ cups white vinegar
3/4 cups water
1 tsp. whole ginger
1 tbsp. whole cloves
3 3-in. cinnaomn sticks

Pour boiling water over peaches and let stand until skins can be easily removed. Dip into cold water and peel. Stick one clove in each peach. Let stand in vinegar-water to prevent browning. Combine sugar, 1½ cups vinegar and 3/4 cup water. Add spices tied in cheesecloth bag. Heat to cooling. Add drained peaches, cover and boil until tender about 10 min. Let stand overnight. Drain, saving liquid, remove spice bag. Pack peaches into clean, hot jars. Bring syrup to boil and pour over peaches, leaving ½ in. depending upon type of lids used. Process 30 min. in boiling water bath. Makes 4 or 5 pt.

Mrs. Val Kuska
Lincoln, Nebraska

Strawberry Pickles

Strawberries
Cinnamon
Cloves
2 cups cider vinegar
3 pt. sugar

Place strawberries in bottom of jar, add layer of cinnamon and cloves, then berries and so on. Pour over it a syrup made of 2 cups cider vinegar and 3 pt. sugar boiled about 5 min. Let stand 24 hr., pour off syrup, boil, pour over berries and let stand as before, then boil berries and syrup slowly for 25 min., put in jars and cover. The above is for 6 qt. of berries.

Mrs. Niels Miller
Herman, Nebraska

Swedish Style Pickled Beets

2 cans (1 lb. size) whole beets
1 cup wine vinegar
½ cup sugar
1 tsp. salt
3 whole cloves
4 small black peppers
1 small bay leaf
2 medium onion (peeled and sliced thin)

Drain beets, reserving 1¼ cups liquid. Place beets in 2 qt. jar or bowl. In saucepan combine vinegar, salt, sugar, cloves, peppers, bay leaf and reserved liquid. Bring to boil. Reduce heat and simmer uncovered 5 min. Add onion slices to beets. Pour hot liquid over beets and onions. Refrigerate, covered several hours. Makes 6 cups.

Mrs. Albert Turner
Loomis, Nebraska

Sweet Beet Pickle Syrup

2 cups brown sugar
2 cups vinegar
1 cup water
½ tsp. salt

For a flavor to each jar add a few seeds of either allspice, caraway seed or anise seed.

Mrs. Elmer Buchholtz
Morse Bluff, Nebraska

Sweet Pickled Beans

Beans
2 pt. vinegar
1 pt. sugar
1 tbsp. mixed spices

Clean and wash beans, then boil in salted water until done, but do not cook to pieces. Place in glass jars and cover with boiling syrup made of 2 pt. vinegar, 1 pt. sugar, 1 tbsp. mixed spices. Sugar, vinegar and spices may be varied in quantity to suit taste. A glass preserve jar about 5½ in. tall, with the white rubber-line lid, is just right for packing the whole beans lengthwise in the jar.

Olga Wangerin
Kensington, Nebraska

Sweet Pickles Made from Dill Pickles (Not Kosher)

½ gal. drained dill pcikles
6 cups sugar
2/3 cup white vinegar
1 tsp. mustard seed
1 tsp. celery seed

Save scant cup of the liquid drained from ½ gal. of dill pickles. Slice pickles crosswise or lengthwise into a crock. Pour pickle liquid over them. Then add 6 cups sugar, 2/3 cup white vinegar, 1 tsp. mustard seed, 1 tsp. celery seed. Stir often until clear. Takes about ½ hr. Makes 4 to 5 pt. Store in refrigerator. Pickles will crisp up in 2 or 3 days.

Mrs. Albert Turner
Loomis, Nebraska

To Convert Dill Pickles to Sweet Pickles

Dill pickles
2 cups vinegar
1 cup sugar
1 or ½ tsp. whole mixed pickling spice

Unseal dill pickles, drain. Rinse thoroughly 2 or 3 times. Drain. Cut lengthwise into smaller pickles. For each qt. of pickles use 2 cups vinegar, 1 cup sugar and 1 or ½ tsp. whole mixed pickling spice. Bring to boiling point. Boil 5 or 10 min. Recan and seal while hot, or let set a day or two before serving. This recipe is good to use when one has run out of sweet pickles but still has some dills.

Mrs. E. B. Staman
Scottsbluff, Nebraska

Uncooked Cucumber Pickles

Good for using up small amount of cucumbers. Good for salads, sandwiches, color poor but taste good (use pints, cutting recipe to fit). Pack hot qt. jar or jars with clean cucumbers. To each qt. add 1 tbsp. crushed rock salt, 1 tbsp. sugar (optional), 1 tbsp. mixed pickling spices. Fill with cold vinegar, seal, store in a cool place. 3-4 slices white onion may be added.

Mrs. Jessie L. Salfrank
Rock Port, Missouri

Unripe Cucumber Pickles (Gherkins)

Cucumbers
1 cup salt
2 qt. water
1 tbsp. powdered alum
1 gal. vinegar
4 red peppers
2 sticks cinnamon
2 tbsp. allspice berries
2 tbsp. cloves

Wipe 4 qt. small unripe cucumbers. Put in stone jar and add 1 cup salt dissolved in 2 qt. boiling water. Let stand 3 days. Drain cucumbers from brine, bring brine to boiling point, pour over cucumbers and again let stand 3 days. Repeat. Drain, wipe cucumbers and pour over 1 gal. boiling water in which 1 tbsp. alum has been dissolved. Let stand then drain from alum water. Cook cucumbers 10 min., a few at a time in 1/4 of the following mixture heated to the boiling point and boiled 10 minutes: 1 gal. vinegar, 4 red peppers, 2 sticks cinnamon, 2 tbsp. allspice berries, 2 tbsp. cloves. Strain remaining liquor over pickles which have been put in stone jar.

Raymond J. Larson
Plattsmouth, Nebraska

Zucchini Pickles

3½-4 lb. zucchini squash
Powdered alum
Water
1½ cups cider vinegar
4 cups sugar
1 tbsp. stick cinnamon
1 tbsp. white mustard seed
2 tsp. whole cloves (tie in bag)

Cut ends off of zucchini squash. Quarter and remove seedy part and soft or pithy center. Do not peel. Cut as you prefer (slices or sticks). Cover with powdered alum and water to firm tissue of squash. 1 tsp. alum to each pt. of water (about 3 pt. water will cover). Let stand overnight. Next morning drain, cover with boiling water and cook 10 min. after it comes to a boil. Drain and cool quickly, let pickles stand in ice water 2 hr. Drain again and use cloths to absorb excess water. Make syrup of 1½ cups cider vinegar, 4 cups sugar, 1 tbsp. stick cinnamon, 1 tbsp. white mustard seed, 2 tsp. whle cloves (tie in bag). Bring spiced syrup to boil and add the squash. Simmer about 1 hr. Pack squash in jars, cover with syrup. Makes about 3 pt.

Chili Sauce

2 qt. peeled ripe tomatoes
2 seeded green peppers
2 peeled onions
1 pt. vinegar
1 tbsp. salt
½ cup brown sugar
1 tsp. cloves
1 tsp. allspice
1 black pepper

Chop tomatoes, peppers and onions and combine and add remaining ingredients. Cook until thick. Adjust jar rings on clean hot jars and fill with hot sauce. Wipe off jar rings and seal while very hot.

Mrs. Ivar G. Saf
Oakland, Nebraska

Tomato Ketchup

Tomatoes
Cider vinegar
White sugar
Salt
Cinnamon
Red peppers
Onions

Cook tomatoes in their own juice until soft, then strain to 1 gal. tomato juice and pulp. Add 1 qt. cider vinegar, 1 pt. white sugar, 1 tsp. salt and 1 tsp. cinnamon. Grind 4 red peppers, 5 large onions, cook these with tomatoes. Rub through colander and cook until thick. Takes 3 to 4 hr. of slow cooking. This should be bright red when done.

Mrs. Ivar G. Saf
Oakland, Nebraska

Tomato Soup

Onions
Celery
Tomatoes
Sugar
Salt
Butter
Flour

Chop 6 onions, 1 bunch celery. Start to cook. When celery is done add 3 qt. tomatoes. Cook until tomatoes are done and put all through a food mill. Add 1 cup sugar, ¼ cup salt, 1 cup butter and 1 cup flour, blended together. Process 10 min., 5 lb.

Mrs. E. C. Volkmer
Talmage, Nebraska

Made in the USA
Monee, IL
04 November 2026

67383155R00109